Developing Mathematical Literacy through Adolescent Literature

Developing Mathematical Literacy through Adolescent Literature

Edited by
Paula Greathouse and Holly Garrett Anthony

ROWMAN & LITTLEFIELD
Lanham • Boulder • New York • London

Published by Rowman & Littlefield
An imprint of The Rowman & Littlefield Publishing Group, Inc.
4501 Forbes Boulevard, Suite 200, Lanham, Maryland 20706
www.rowman.com

86-90 Paul Street, London EC2A 4NE, United Kingdom

Copyright © 2022 by Paula Greathouse and Holly Garrett Anthony

All rights reserved. No part of this book may be reproduced in any form or by any electronic or mechanical means, including information storage and retrieval systems, without written permission from the publisher, except by a reviewer who may quote passages in a review.

British Library Cataloguing in Publication Information Available

Library of Congress Cataloging-in-Publication Data

Names: Greathouse, Paula, editor. | Anthony, Holly Garrett, 1976–, editor.
Title: Developing mathematical literacy through adolescent literature / edited by Paula Greathouse, Holly Anthony.
Description: Lanham : Rowman & Littlefield, [2022] | Includes index. | Audience: Ages 12–18 | Audience: Grades 7–9 | Summary: "Students are offered opportunities to explore multiple mathematical topics such as probabilities, statistics, linear equations, integers, and sequencing, as well as algebra, pre-calculus and calculus concepts through literature. As students develop mathematical literacy, they will also explore literary elements such as characterization, setting, and conflict"—Provided by publisher.
Identifiers: LCCN 2021041974 (print) | LCCN 2021041975 (ebook) | ISBN 9781475861525 (cloth) | ISBN 9781475861532 (paperback) | ISBN 9781475861549 (epub)
Subjects: LCSH: Mathematics—Study and teaching (Secondary)—Juvenile literature. | Mathematics in literature—Juvenile literature.
Classification: LCC QA19.L58 D73 2022 (print) | LCC QA19.L58 (ebook) | DDC 510.71/2—dc23
LC record available at https://lccn.loc.gov/2021041974
LC ebook record available at https://lccn.loc.gov/2021041975

About the *Adolescent Literature as a Complement to the Content Area* Book Series

The need to develop students' literacy skills and practices in tandem with content has been a growing concern for many teachers. Teachers of all content areas are now required to address students' literacy development within their discipline—what was once believed to be the responsibility of the English teacher has become the responsibility of *all* teachers. When faced with this expectation, content area teachers often ask how they can do both. Our answer—through the inclusion of adolescent literature in the content area classroom.

Adolescent literature can hold the key to unlocking the common rift between content and literacy-driven instruction in content area classrooms. However, the question most often asked is "How can today's content teachers infuse narrative texts into an already overwhelming curriculum?" The *Adolescent Literature as a Complement to the Content Area* book series provides the answer. Coedited by literacy and content area experts, this series spotlights the secondary (grades 6–12) core content areas of mathematics, science, and social studies/history. Each volume is devoted to one content area and each chapter within that volume spotlights specific content concepts and offers practical approaches for teaching these through a single young adult novel. Experts and educators in each content area field share before, during, and after reading instructional activities, as well as extension activities that move beyond the text. In many cases, the activities presented in the chapters build on each other, and in other cases, they exist independently, allowing teachers to pick and choose those that fit their students best.

In our standards-driven classrooms, making concepts relevant and meaningful to students while meeting the literacy and content requirements of our curriculum can be a challenging task. However, it is a task that can be easily accomplished. Reading adolescent literature in content classrooms provides access points for students to practice reading like content area experts. Giving students opportunities to read this way as they explore content has the potential to move their thinking and understandings in monumental ways.

Contents

Introduction   ix
Paula Greathouse and Holly Garrett Anthony

1. The Mathematics of Imagination: Uncovering Math and Meaning in the Graphic Novel *The Magic Fish*   1
   Katherine Baker, Summer Melody Pennell and Bryan Fede

2. Take It to *The Limit:* Maximizing Mathematical Literacy and (Re)solving Conflict   25
   Holly Garrett Anthony, Paula Greathouse and Kaylee Gentry

3. Extracting Mathematical Topics Embedded in *Holes* and Examining the Text through a Critical Lens   43
   Marilyn E. Strutchens, Mike P. Cook and Brea C. Ratliff

4. Reading like a Mathematician: Exploring Data, Narratives, and Phenomena in *A Long Walk to Water*   63
   Suki Jones Mozenter and Robin Keturah Anderson

5. Exploring *All of the Above:* Platonic Solids, Scale, Proportions, and Characterization   79
   Jennifer R. Meadows and Amber Spears

6. Reasoning with Equations and Functions in *The 5th Wave*   99
   Rebecca Grice Gault and Jennifer Edelman

7. "Absolute Zeros Solve for Why" Using the Pythagorean Theorem in *Island of the Unknowns*   117
   Brian Rothbaum and Julie Grasfield Weil

8 Exploring Math, Culture and Stories through *Math Girls* 131
Shelly Shaffer and Carlos Castillo-Garsow

9 Making a Deal with *The Number Devil:* Scaffolding, Deepening, and Extending Students' Mathematical Literacy 155
Amanda Huffman, Rachel Colby, Jenna Repkin and Shelly Furuness

10 Matchmaking Mathematics: Teaching Algorithms and Probability with Nandini Bajpai's *A Match Made in Mehendi* 175
Jen McConnel and Allen Harbaugh

11 Simulating Success Based on Societal Odds: A Mathematical Read of *The Hunger Games* 191
Robin Keturah Anderson, Melissa Troudt, Candace Joswick, and Lisa Skultety

Subject Index 211

About the Editors 213

About the Contributors 215

Introduction

Paula Greathouse and Holly Garrett Anthony

In our standards-driven courses, making concepts relevant and meaningful to students while meeting the requirements of the curriculum can be a challenging task. Educators often rely on district- or school-assigned textbooks and resources publishers create as supplements to determine what needs to be learned and how (Budiansky, 2001; Daniels & Zemelman, 2004). But what is presented in these mediums is often far removed from the lives of the students who are expected to learn from them. An unintended consequence is the wavering of engagement and motivation to learn, as students see no relevance of the content to their lives (Moore, Alvermann, & Hinchman, 2000). It has been posited that instead of requiring students to learn isolated facts presented in textbooks, courses should be designed to connect major concepts to one's world (Levine et al., 2008). In doing so, there is potential to increase student interest and engagement in learning, as well as an opportunity to provide students with a more tangible idea of concepts in the "real-world" (Larson & Rumsey, 2017; Newhouse, 2018; Price & Lennon, 2009). One way educators can accomplish this is through the inclusion of young adult literature (YAL) in their content area courses. YAL features characters, plots, and themes that are relevant to the lives of students. As such, pairing YAL titles with concepts being learned through textbook readings can promote greater connection and understanding of content topics.

Specific to mathematics, including YAL as a complement to the curriculum can extend students' understanding of complex mathematical concepts (Austin, Thompson, & Beckmann, 2005; Bean, 2003). The inclusion of YAL in the mathematics classroom can introduce students to new vocabulary, help them make connections among abstract concepts, and offer them a way to see mathematics across content areas (Whitin & Wilde, 1992). When YAL is read as a complement to the mathematics textbook, students learn to synthesize concepts across a range of texts (Bean, 2003) while providing focus and coherence to content area instruction (Austin, Thompson, & Beckmann, 2005). As such, reading YAL in mathematics classrooms can provide access points for students to practice reading like content area experts—mathematicians. Giving students opportunities to read like mathematicians as they explore content has the potential to move their thinking and understandings in monumental ways. In mathematics classrooms, students can be encouraged to draw on imagination as they solve problems presented in the text as a way to consider the possibilities of mathematics in their world and future (Koellner, Wallace, & Swackhamer, 2009).

THE COLLECTION—PURPOSE AND ORGANIZATION

In our conversations with mathematics teachers, we discovered that very few have even considered utilizing YAL in the mathematics classroom. The reason for this has been attributed to their lack of knowledge and exposure to YAL as well as guided instructional strategies they can implement using these texts in their curriculum. In other words, mathematics teachers are unclear on ways in which they can incorporate YAL into their courses as a medium through which to engage students in dialogues about content to develop mathematical literacy. Our book seeks to remedy this by offering educators a collection of practical approaches for teaching mathematical concepts through YAL.

Throughout this collection, authors use the term "secondary" as opposed to middle school or high school; this is intentional as even though some YAL is often associated with specific grade levels, we have known many teachers who have used these texts across all grade levels. We also refrain from specifying specific reading levels for any of the texts discussed as we have found assigning Lexile and grade levels to be restrictive without adding any benefits. We intentionally leave those types of decisions to our teacher readers who know their students best. Similarly, although the pedagogical approaches offered within the chapters align with current mathematics, English language arts, and literacy standards, we eschewed referencing any specific ones as lists of standards can become unwieldy in texts and because teachers can easily determine how activities meet their local requirements.

Each of the chapters presented in this volume are organized correspondingly, with an introductory section, a summary of the text, and then instructional activities for before, during, and after reading; furthermore, each chapter includes extension activities that move beyond the text. In many cases, activities build on each other, and in other cases, they exist independently, allowing teachers to pick and choose those that fit their students best. Because our collection is geared toward secondary teachers, we have organized it in the order in which mathematical concepts are often taught in secondary schools, acknowledging that this ordering is not definitive.

Our collection opens with Katherine Baker, Summer Melody Pennell, and Bryan Fede's exploration of rational numbers in "The Mathematics of Imagination: Uncovering Math and Meaning in the Graphic Novel *The Magic Fish*." Their chapter outlines strategies for teaching about rational numbers, such as the density principle and identifying and reasoning about fractions, alongside English language arts concepts of visual literacy and author's use of time through the reading of the young adult (YA) graphic novel *The Magic Fish* (Nguyen, 2020).

Holly Garrett Anthony, Paula Greathouse, and Kaylee Gentry spotlight mental mathematics, integers, averages, and debt through *The Limit* in "Take It to *The Limit*: Maximizing Mathematical Literacy and (Re)solving Conflict." Combined with literary analysis of conflict, approaches for students to analyze these mathematical concepts include activities that also develop student's financial literacy.

Through the YA novel *Holes* (Sachar, 2000), Marilyn E. Strutchens, Mike P. Cook, and Brea C. Ratliff's chapter, "Extracting Mathematical Topics Embedded in *Holes* and Examining the Text through a Critical Lens," offers approaches for students to explore the mathematical concepts of probability, statistics, and proportional reasoning. In addition, approaches for developing mathematical problem-solving agency and sociopolitical agency are shared as a way to help students interrogate the interrelationship between, and the inseparability of, literacy and power.

In "Reading like a Mathematician: Exploring Data, Narratives, and Phenomena in *A Long Walk to Water*," Suki Jones Mozenter and Robin Keturah Anderson share approaches that engage students in a study of mathematical concepts of scale, perspective, and extent through the reading of *A Long Walk to Water* (Park, 2010). In this chapter, they also provide activities that afford students an opportunity to connect issues relating to refugee experiences and safe water access to their local context via mathematical practices.

Jennifer R. Meadows and Amber Spears's chapter, "Exploring *All of the Above*: Platonic Solids, Scale, Proportions, and Characterization," offers readers approaches for guiding students through surface area of three-dimensional figures using two-dimensional nets, as well as using proportional reasoning to solve problems as they read *All of the Above* (Pearsall, 2006).

Through *The 5th Wave*, Rebecca Grice Gault and Jennifer Edelman's chapter, "Reasoning with Equations and Functions in *The 5th Wave*," covers topics typically included during an algebra course including creating equations, reasoning with equations and inequalities, interpreting functions, and building functions. The literacy focus is the study of dystopia as a genre and includes the development of critical lenses through which students examine literature and, by extension, culture.

In "'Absolute Zeros Solve for Why' Using the Pythagorean Theorem in *Island of the Unknowns*," Brian Rothbaum and Julie Grasfield Weil share approaches for developing students' understanding of the Cartesian coordinate plane, Pythagorean Theorem, and probability. As students read *Island of the Unknowns* (Carey, 2009), the authors also provide activities that guide students through an analysis of plot and structure.

Shelly Shaffer and Carlos Castillo-Garsow encourage students to engage with precalculus concepts in their chapter "Exploring Math, Culture, and Stories through *Math Girls*." Through the reading of *Math Girls* (Yuki, 2011), students study characters' interactions with mathematics and how they engage in key mathematical processes and proficiencies by analyzing characters' motivation to complete mathematics problems, persevering to complete difficult problems, providing feedback to one another, and constructing viable arguments.

Amanda Huffman, Rachel Colby, Jenna Repkin, and Shelly Furuness describe activities for students to engage in the study of primes, patterns, and series through the reading of *The Number Devil* (Enzensberger, 1997) in "Making a Deal with *The Number Devil*: Scaffolding, Deepening, and Extending Students' Mathematical Literacy." Their chapter includes activities that ask students to collaborate, create, and organize thoughts and understanding of these mathematical concepts in addition to the crafting of a mathematical dream journal in which the prompts call for concept synthesis.

Jen McConnel and Allen Harbaugh's chapter, "Matchmaking Mathematics: Teaching Algorithms and Probability with Nandini Bajpai's *A Match Made in Mehendi*," shares ways in which Nandini Bajpai's contemporary YA novel can be integrated into conversations about algorithms, probability, and variables by focusing on the following driving questions: Which variables affect authentic relationships? And, how can you measure and plan for these variables?

In our final chapter, "Simulating Success Based on Societal Odds: A Mathematical Read of *The Hunger Games*," Robin Keturah Anderson, Melissa Troudt, Candace Joswick, and Lisa Skultety present ideas for integrating mathematics into the reading of *The Hunger Games* (Collins, 2010) to promote a critical analysis of an individual's odds of success. Students use mathematics to investigate odds and theoretical and experimental probabilities of both independent and dependent events and discover that the novel's classic mantra "May the odds be ever in your favor" is often out of the individual's control.

REFERENCES

Austin, R. A., Thompson, D. R., & Beckmann, C. E. (2005). Exploring measurement concepts through literature: Natural links across disciplines. *Mathematics Teaching in the Middle School, 10*, 218–224.

Bajpai, N. (2019). *A match made in Mehendi*. Little, Brown and Company.

Bean, T. W. (2003). *Using young-adult literature to enhance comprehension in the content areas*. North Central Regional Educational Laboratory.

Budiansky, S. (2001). The trouble with textbooks. *Prism Online, 10*, 24–27.

Carey, B. (2009). *Island of the unknowns*. Amulet Books.

Collins, S. (2010). *The hunger games*. Scholastic Press.

Daniels, H., & Zemelman, S. (2004). Out with textbooks, in with learning. *Educational Leadership, 61*(4), 36–41.

Enzensberger, H. M. (1997). *The number devil: A mathematical adventure*. Picador.

Koellner, K., Wallace, F. H., & Swackhamer, L. (2009). Integrating literature to support mathematics learning in middle school. *Middle School Journal, 41*(2), 30–39.

Landon, K. (2010). *The limit*. Aladdin.

Larson, L. C., & Rumsey, C. (2017). Bringing stories to life: Integrating literature and math manipulatives. *The Reading Teacher, 71*(5), 589–596.

Levine, L. E., Fallahi, C. R., Nicoll-Senft, J. M., Tessier, J. T., Watson, C. L., & Wood, R. M. (2008). Creating significant learning experiences across disciplines. *College Teaching, 56*(4), 247–254.

Moore, D. W., Alvermann, D. E., & Hinchman, K. A. (2000). *Struggling adolescent readers: A collection of teaching strategies*. International Reading Association.

Newhouse, K. (2018, March 18). *How reading novels in math class can strengthen student engagement*. https://www.kqed.org/mindshift/50640/how-reading-novels-in-math-class-can-strengthen-student-engagement.

Nguyen, T. L. (2020). *The magic fish*. Random House Graphic.

Park, L. S. (2011). *A long walk to water*. Houghton Mifflin.

Pearsall, S. (2008). *All of the above*. Little, Brown Books for Young Readers.

Price, R., & Lennon, C. (2009). *Using children's literature to teach mathematics*. Quantile.

Sachar, L. (2000). *Holes*. Yearling.

Whitin, D. J., & Wilde, S. (1992). *Read any good math lately? Children's books for mathematical learning, K–6*. Heinemann.

Yancey, R. (2013). *The 5th wave*. G.P. Putnam's Sons.

Yuki, H. (2011). *Math girls*. Bento Books, Inc.

Chapter 1

The Mathematics of Imagination

Uncovering Math and Meaning in the Graphic Novel The Magic Fish

Katherine Baker, Summer Melody Pennell and Bryan Fede

While mathematics topics may not be overtly stated within its text, *The Magic Fish* (Nguyen, 2020) embodies mathematical sensemaking in its creation as a graphic novel. What to draw, how to capture social algorithms, how many panels should be on each page, and where and why to place particular drawings and panels in relation to others are all mathematical decisions. Our collaboration as authors and this chapter is grounded in the work of humanizing mathematics (Gutiérrez, 2010; Yeh & Otis, 2019) and queering mathematics (Pennell & Fede, 2017) that position mathematics as ubiquitous in everyday life and promotes the questioning of mathematics that would otherwise be tacitly accepted as "fact" (Meyer, 2019).

Traditional mathematics knowledge taught in schools buys into the idea of mathematics as facts, and as represented through textbooks, curriculums, and assessments, whereas a student's funds of knowledge recognize the community-based, inherent, intuitive mathematical thinking students bring to school (González et al., 2001). We recognize mathematics as a human endeavor representing humans' lived experiences and identities, and that this endeavor is contextualized in sociocultural, political, and power-based issues. When we queer mathematics, we invite a multitude of ideas and strategies into the mathematics space, and trouble the notion of one direct line of thinking or one neat answer to a problem. We instead recognize that part of the beauty of experiencing mathematics is that it can offer robust discussion, alternate ways of thinking, and creative ways of solving problems surrounding us. Secondary students are already experiencing mathematics simply through being, and a graphic novel is mathematics in its existence as art and as social commentary, thus making *The Magic Fish* an ideal fit for generating mathematical conversation with multiple viewpoints.

Integrating this graphic novel with mathematics content is best done when creating a space that acknowledges humanizing mathematics values and encourages student-led sensemaking and discussion. We use *The Magic Fish* to set the stage for bringing students' funds of knowledge into the mathematics classroom space, inviting discussion, multiple solution ideas and strategies, and centering the unpacking of conceptual understanding about rational numbers. When talking about rational numbers, we follow the definition given by Beckmann (2018): "Rational numbers are those numbers that can be expressed as a fraction or the negative of a fraction" (p. 363). A more formal definition of rational numbers is any number that can be expressed in the form *a/b* where *a* and *b* are both integers and *b* is not zero. In this chapter we offer suggestions for exploring how investigating rational numbers of all types (fractions, decimals, or whole numbers) can deepen literary analysis of themes and how the author's choices create meaning in *The Magic Fish*.

THE MAGIC FISH BY TRUNG LE NGUYEN

Nguyen's (he/they pronouns) graphic novel is a layered story of a boy, Tiến, whose parents immigrated to the United States from Vietnam before he was born. He reads fairy tales aloud with his parents so they can practice their English, and these serve as extended metaphors for themes of finding yourself, finding support, and changing your own story. It is also a coming-out story, as Tiến is trying to navigate telling his parents he is gay without knowing how to talk about that in Vietnamese. Coming out is even more complicated for him because he is attending a Catholic school in the 1990s (in one scene, there is a news story about Matthew Shepard on the television). This novel concludes with Tiến's mother, Hiền, showing unconditional support for his identity, and reinforces the power we have to create our own story. Nguyen was both the author and illustrator, and the ending author's note gives insight into his creative process.

BEFORE READING *THE MAGIC FISH*

In order to create a classroom where students' funds of knowledge are honored, teachers need to create a space where students' experiences and their mathematical thoughts, mathematical contributions, and informal mathematical ideas are elicited and valued. Students need to be ready to engage in productive discussions in order to engage in the mathematical work of *The Magic Fish*. The National Council of Teachers of Mathematics (2014) charges teachers with facilitating meaningful mathematical discourse among students to build shared understanding of mathematical ideas. It also charges teachers to support their students' individual and collective productive struggle by allowing them to grapple with mathematical ideas. Teachers and students need to practice engaging in mathematical discussions around open-ended mathematics tasks where the answers are not immediately apparent. Chapin and colleagues (2013) offered support in mathematics discussion facilitation, and specifically shared talk moves that teachers can use to encourage student discussion. These talk moves include teacher revoicing (repeating some or all of the information shared by a student and allowing the student to confirm or edit), students restating (students repeat or rephrase the reasoning shared by a classmate), and teacher questioning techniques that press for reasoning ("How did you figure that out?" or "What's your evidence?").

Preparing Students for the Mathematics

In order to prepare students for the rational number work in *The Magic Fish*, teachers could pose a mathematical inquiry around the following prompt: Identify any number between 1 and 3. Defend your answer with a representation. Mathematically undergirding this prompt is the idea that between any two numbers are an infinite amount of other numbers. This prompt is intentionally open-ended and vague to begin with to encourage robust discussion and productive struggle. It may help the discussion if students are familiar with number lines and how to use number lines to "zoom in" to find other values (see figure 1.1).

In the image that follows are four number lines that purposefully do not have a common scale to highlight how there is an infinite amount of numbers between any two numbers. The image is broken down into four phases to show how you could find a number like 1.576 between 1 and 2, and better articulates the relationship of whole numbers and decimal numbers. The Phase 1 number line marks whole numbers between 0 and 10 with 10ths represented between each number. The Phase 2 number line zooms in between 1 and 2, as denoted by the shading, and marks the 10ths between 1 and 2. Between each of the 10ths are spaces of 100ths which then leads to the Phase 3 number line. This Phase 3 number line zooms in between 1.5 and 1.6 marking the 100ths that fall between the two numbers and showing 1000ths between each hundredth. This leads to the Phase 4 zooming between 1.57 and 1.58, marking the 1000ths and showing that you could find 1.576 between the original 1 and 2. The Phase 4 number line could then be zoomed into 10,000ths, and so on, emphasizing the infinity property of numbers.

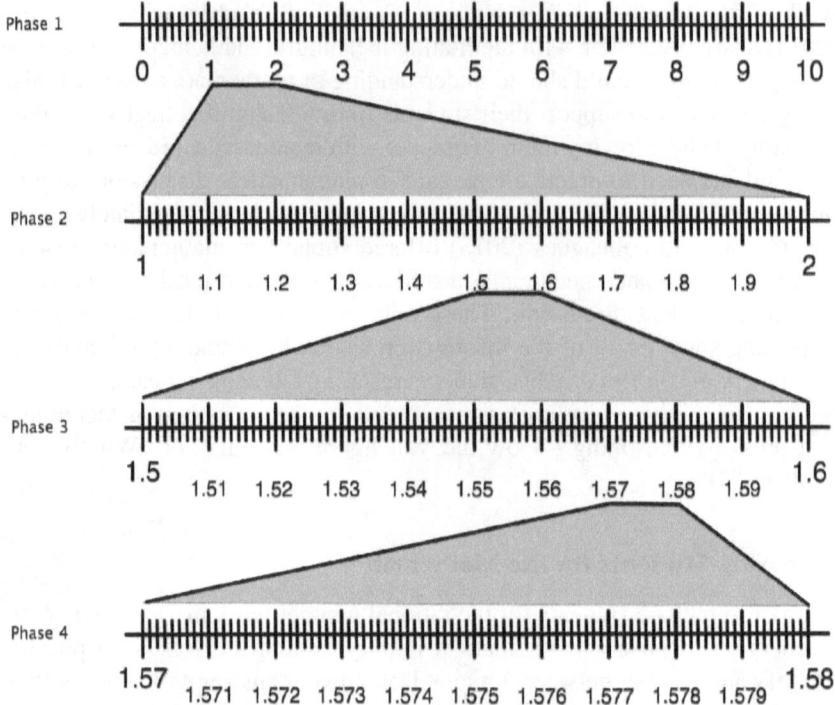

Figure 1.1 Zooming in on the Number Line to Find the Number 1.576. *Source:* Adapted from Beckmann, 2018

Part of a conversation about the prompt is provided next to show how it may begin to play out in the classroom space. Note that the discussion starts with open-ended language of "between" 1 and 3 and that students might interpret this as the middle. However, as the guiding mathematical parameters from the teacher become more detailed, students will start to see that while their responses might all be valid, they may also need to be adjusted to reflect the changing parameters. The novelty of the zooming-in activity on the number line highlights a mathematically amazing concept for students—that between any two numbers, there is an infinite amount of other numbers. Like the name of this chapter, students may soon start to see their limits on finding and identifying numbers between numbers are limited only to their imaginations.

Teacher: I see that some of us answered and defended 2, and then there is another answer of one-and-a-half. What do we think? Can there be more than one right answer for this question? Let's vote with a thumbs up or thumbs down. (Students vote and most give a thumbs up). I see that a lot of us are thinking there could be different, valid answers. So, let's dig into one-and-a-half then. Why might one and a half work?

Student A: One and a half is also between 1 and 3, it's just not a whole number. See (shows number line) it still falls between 1 and 3, just in the middle of 1 and 2.

Student B: Wouldn't that be in the middle of 1 and 2 instead? So it doesn't fit the question?

Teacher: What do we think, does one and a half fit the question?

Student A: Yes, because it's still between 1 and 3, and you didn't say where between 1 and 3 we had to be. So, it doesn't have to be a whole number.

Teacher: So if it's not a whole number, what other numbers could it be?

Student C: Like fractions or decimals?

Teacher: Lets dig into that more and change our question slightly to say that we have to identify a number between 1 and 3 that involves a fraction or decimal. How might that change your answer and your representation? Work on revising your answer individually then we'll share our work in partnerships and discuss our choice.

Using this prompt serves several purposes. For teachers, it encourages practicing facilitating discussions and making connections between student strategies and mathematical arguments, encouraging opening the space to the queering and humanizing values we share in the opening. It also serves as a formative assessment to hear how students think about fractional and decimal values, especially those with odd numbers in the denominators. For students, they get practice thinking through problems without immediately apparent answers, and they see that problems can have multiple accurate answers, again establishing queer math ideals. They will also practice defending their thinking.

Preparing Students for Reading a Graphic Novel

In order to prepare students for how to read and interact with a graphic novel, students will need visual literacy instruction. Guide students through an exploration of a few pages from a graphic novel so they can see that you read the panels from left to right and top to bottom. Provide students with the definitions in the student handout (figure 1.2), and have them identify these elements within *The Magic Fish* or a sample comic page for practice. Any comic will work, but asking students to bring in a comic they enjoy reading would increase engagement. A single sample page is unlikely to have all elements, so teachers could assign students different sections of a graphic novel or comic and encourage students to record what they find, to share with their classmates later. Along with identifying features, students could also discuss

Element	Definition	Page Examples from the Novel
Panel	Boxes or other shapes that contain pictures and words. There can be several panels on a single page. Their size and shape can indicate meaning and importance.	1 (found on most pages)
Gutters	The space between panels.	1 (found on most pages)
Word balloons/ Speech bubbles	Text that shows speech. The tail points to the speaker. The text may be bold and upper case to show shouting, or small and light to show a whisper.	3
Thought bubble	Bubble is drawn with curvy edges like a cloud, and has bubbles leading to the thinker, indicating the text is a character's thoughts and not speech.	58
Captions	Words in a box; usually give narration.	1
Sound effects	Words such as "wham!" may be written inside jagged bubbles to indicate sounds. The shape of the bubble, and the weight of the text, also indicate meaning.	113
Motion lines	Marks that indicate a character or object's movement.	25
Lighting	The way light is drawn can indicate time of day, but also mood. For example, darker panels may indicate something scary, or mysterious.	18
Color	Used to indicate visual interest as well as meaning. This could include mood, characterization, a sense of time, etc.	1

Figure 1.2 **Graphic Novel Elements Handout.** *Source*: From Ankiel, n.d.; Pavlik, 2021; Reese, n.d.

how these features create meaning. In other words, what do they learn about the text from these visual components? Students should use this sheet for reference throughout discussion of the novel so that they can refer to the appropriate terms when discussing the author's craft, theme, and when using components for mathematical problems.

Cultural Understandings

For students to fully understand Tiến's family in *The Magic Fish*, some background information on Vietnamese culture, immigration from Vietnam to the United States, and the impact of having immigrant parents can be explored. Students could do research in their English Language Arts (ELA) class (or with support from an ELA teacher in a mathematics class) in small groups as a jigsaw, with each group investigating one of these questions: *(1) Why*

Table 1.1 Suggested Sources on Immigration

Suggested Sources	Source Description
National Geographic: Resettling Vietnamese Refugees in the United States [website]	Resource library including video, text, and vocabulary. Designed for students grades 6+.
Asian Americans Advancing Justice: SE Asian Refugees [website]	Video & vocabulary (also includes full lesson plan for teachers)
Learning for Justice: Immigration [website]	Topic page includes links to sources such as "10 myths about immigration" and narratives about immigration.
Growing up in American with Immigrant Parents [essay on Writing our Future: American Creed website]	Personal essay written by a high school student. Website is a project by the National Writing Project.
The Immigrant Learning Center [website]	Provides research reports, infographics, and other data. You can search by country or topic. Reading level may be difficult for younger students.

did Vietnamese people immigrate to the United States in the 1970s? (2) What does it mean to be a refugee? (3) What struggles do immigrants to the United States face once they get here? (4) What are some experiences of kids whose parents are immigrants? After students have researched their question, they can share their findings with their heterogeneous group so that all students learn about each topic. See Table 1.1 for suggested resources, and consider asking your school librarian for additional sources. Teachers can create a graphic organizer using these questions so that all students can use this as a reference guide while they read the novel.

The Impact of Storytelling and Fairy Tales across Cultures

To prompt students to consider a major theme in the novel, teachers can engage students in a discussion about the impact of storytelling in their lives. Nguyen uses three fairy tales as metaphors for characters' experiences, and to highlight that stories can be used to communicate common themes across cultures. These stories and the corresponding page numbers are listed in Table 1.2 for teacher reference. The book also demonstrates that we can change our own narratives to create our happy endings depending on our circumstances. As Hiền narrates, "Fairy tales . . . can change [. . . and] I imagine the [basic story] stays the same, but the context always shifts" (Nguyen, 2020, p. 5). These ideas could be discussed in small groups or as a whole class, and done

Table 1.2 Fairy Tales in *The Magic Fish*

Fairy Tale	Nguyen's Inspiration (in author's note)	Page Numbers
Tattercoats	From the German "Alleirauh"	1, 8–26, 28–40, 50–64, 80–82, 85, 92–112
Cinderella	From the Vietnamese "Tấm Cám"	2, 132–147, 151–154, 157–158, 160–165, 168–170, 174–175, 178-183
The Little Mermaid	From Hans Christian Andersen's version, illustrated from Hiền's "visual imagination" (Nguyen, 2020, n.p.)	2, 187–213, 218–227

orally or as journaling prompts. The following are the suggested questions: *(1) What can we learn about ourselves and our communities from fairy tales and other stories? (2) How can stories communicate complicated ideas and feelings? (3) How do stories connect us to our home(s)? (4) How can we change our own stories as our contexts shift?* As students read, continue discussing the context in which the stories are told and how that affects their meaning for the characters.

Since *The Magic Fish* explores different versions of Cinderella (from German and Vietnamese cultures), students could research this tale to learn how it compares across groups and regions. Table 1.3 has suggestions for finding different versions, and teachers may want to consult with their school librarian to find print versions. After conducting research, discuss how many versions of Cinderella were explored and their differences. *What changes do you notice from the version you are familiar with? What do those changes tell you about the context of the story?* Teachers can integrate mathematics by posing the following question: *How many of those versions' geographic locations are represented in the school community?* Students could survey their classmates and school staff to investigate this question, and also survey their classmates, family members, and friends to find out what fairy tales they know best. They can then compare the data and identify fractions for their own data set and/or collectively across the class set, answering questions such as the following: *How many people like the same fairy tale? How many fairy tales are from Eastern cultures? How many fairy tales are from Western cultures?*

Table 1.3 Other Versions of Cinderella

Suggested Source	Description
"Cinderella" Folk Tales: Variations in Plot and Setting [website]	This lesson plan from EDSITEment (from the NEH) includes a video of a read aloud of "Rhodopis" the Egyptian version of Cinderella
Cinderella Tales: 10 International Versions of the Beloved Tale [website]	Compiled on the Fairytalez website, brief descriptions are provided with links to read the full tale.
Multicultural Cinderella Stories [website]	A curated book list by Mary Northrup for the American Library Association (ALA). Books are divided by continent, with an additional list of parodies. Brief descriptions of the books are included.

WHILE READING *THE MAGIC FISH*

When first launching into reading Nguyen's text, teachers can offer open-ended discussion points that welcome students' considerations about the mathematics represented by a graphic novel's format and images, such as the following: *What math are you noticing on the cover of this book? When you flip through the pages?* Record students' ideas and leave them in a common digital or shared classroom space to refer to and add to as reading occurs. Journaling and turn-and-talks about the mathematics topics students are noticing could be included throughout the reading of the entire text. We encourage teachers to be genuinely curious about the mathematical moments their students are identifying, and try to follow these paths to meet mathematical goals. We offer a suggested reading calendar in Figure 1.3 and specific mathematics explorations for the first six reading calendar sections, and a review or extension suggestions for the final sections. We also offer ongoing strategies for literary analysis that can encompass the entirety of the graphic novel.

Exploring the Mathematics

In *The Magic Fish* there are mathematical undertones that serve as interesting launch points for mathematical investigations. Teachers can use the reading of *The Magic Fish* to establish a shared classroom context for the exploration of and sensemaking around the meaning of rational numbers, as well as invitations into how rational numbers interact (comparison between,

Section	Pages	Math and ELA Possibilities
1	1–23	• Math: Exploring infinity and fractional reasoning through a zooming-in investigation of panels and color use. • ELA: Visual literacy- color use and meaning. Importance of language for immigrant families.
2	24–40	• Math: Naming rational numbers as fractions and constructing equivalent fractions. Tracking the use of color to distinguish between the storylines. • ELA: Connections between narratives and cultures. Flashbacks.
3	41–64	• Math: Reasoning about the relative size of fractions and name them in relation to a changing whole. • ELA: Visual literacy- panel and word bubble structure and placement
4	64–85	• Math: Exploring how fractions change as wholes get bigger and smaller and discussing fractional relationships, like comparison of fractional amounts between storylines. (Note: This reading section begins with page 64 again to allow for a mathematics exploration across pages 64–65.) • ELA: Connections between narratives; analyzing panels without text; flashbacks.
5	86–116	• Math: Noticing and wondering about rational numbers and zooming-in approach revisited; connect mathematics back to the zooming work in "Before Reading" and Section 1. • ELA: Connections between "Tattercoats" and Tiến's family
6	117–131	• Math: Incorporating identity and social themes in mathematics, spurred by Tiến's mother flying home on pages 120–121. • ELA: Visual literacy- expressions of emotion and Hiền's travel.
7	132–154	• ELA: Visual literacy- analyzing drawings to examine relationships between characters. Fairy tales across cultures.
8	155–177	• ELA: Visual literacy- analyzing how the author shows the fairy tale and present day are connected. Juxtaposition of fairy tales with Tiến's narrative. Reactions to Tiến coming out. Difficulties in cross-cultural communication.
9	178–206	• ELA: Fairy tales, visual cultural cues, and narrative connections- The Little Mermaid
10	207–229	• ELA: Narrative structure and changing narratives. Themes of unconditional love and support, cross-cultural communication for children of immigrants.

For sections 7–10:
• Math: Revisit and reinforce the math topics of sections 1-6 by applying the ideas and skills to the reading of these new sections. For example, *compare the overall prevalence of storylines throughout by finding the total space dedicated to each storyline in the novel as a whole.*
• Or, Extend into some of the geometry and measurement work needed for the "After Reading" investigation based on the structure of the graphic novel. Students could solve measurement problems, such as measuring for and calculating the area of a page and the areas of its various panels.

Figure 1.3 **Suggest Reading Calendar.** *Source*: Created by authors

operations with, etc.). The classroom community formed with the investigations becomes a humanizing space, validating that there are many paths to solutions and many valid solutions. Others' opinions, ideas, and contributions must be listened to, responded to respectively even in disagreement, and students' voices must be uplifted as co-teachers.

Section 1 (Pages 1–23): Zooming-in on the Number Line

At the opening of the book, *The Magic Fish* presents us with an interesting thought experiment: "They say that we're meant to go from here to there, but so much happens between those two places. . . . And there's always more, isn't there?" (p. 1). In this first mathematics investigation, we invite students to model the concept of infinity by visualizing the space between numbers on a number line. It is likely that students will have at least superficial or textbook experiences with decimals and fractions prior to this experience, and will have some foundational discussion thoughts around rational numbers if they dove into the possibilities of the "Before Reading" prompt. However, this section will encourage them to more deeply explore everything that is between two whole numbers, and consider why knowing this might be useful and applicable knowledge. Extending from the opening prompt, now ask students to identify as many numbers as possible between 1 and 2. If the zooming-in on a number line approach (see figure 1.1) has not yet been introduced to students, this would be the time to introduce and explore that model collaboratively. *What does the zooming-in approach afford? Can you zoom in past 100ths? Past 1000ths? Still smaller? How small?* At the heart of this investigation is the revelation that between any two real numbers, there exists another real number, an idea that mathematicians refer to as the density principle. The zooming-in investigation is also a means for discussing the Base 10 number system and its efficiency and purpose. We are able to zoom in by powers of 10 from the 10ths, 100ths, and 1000ths because we are operating in a Base 10 system.

Section 2 (Pages 24–40): Naming Fractions by Exploring Panels and Use of Color

In addition to using *The Magic Fish* as a basis for number line explorations, we want to acknowledge the book as being beautifully illustrated. It is likely that it will be the illustrations and interwoven story lines that will really capture the imagination of students. These illustrations will also have a deeper purpose as teachers and students continue to explore rational numbers. Nguyen used color—red, blue, yellow—as a visual scaffold to assist the reader in keeping track of the various story lines, and students might want to keep track of these color changes noting pages of significant color changes (see Table 1.4; color tracking also discussed in the "Ongoing Literary Analysis" section). Pages feature a different proportion of these colors to emphasize a story being told and woven together. As the stories change, the proportions change. In Nguyen's craft, the illustrations and colors support readers to seamlessly jump from one story line to the next—and even back again—all within one page in some cases.

Table 1.4 Tracking Colors in *The Magic Fish*

Implementation Note: Teachers can decide if they want to give students the option of representing proportions of colors in fractions, decimals, or both. Here fractions are used, as in our examples, to show the proportion of color on the total page. Teachers may want to add an additional column to track when multiple colors appear in a single panel, or when colors blend together, as discussed in the "ongoing literary analysis" section.

Page	Red	Yellow	Blue	Meaning
1	1/4	1/4	1/2	The blue portion is the biggest, so it must be the most important to this page. The drawings are also more detailed in this section.
28	1/9	5/9	3/9	The focus is on the blue (fairy tale) and yellow (flashbacks). The blue panels cut through the yellow ones diagonally, showing that Hiền is remembering her scary journey from Vietnam by boat while reading about the scary Old Man in the Sea in "Tattercoats." The 1 red panel shows us she is thinking of both of these while reading with Tiến.
29	1/3	0	2/3	The red panels bring us to the present, and show Hiền hiding her stressful memory from her son. The blue panels take us back to "Tattercoats." The bigger size of the panels (compared to p. 28) shows that Hiền is purposefully focused more on the story and talking with Tiến than on her past, as each storyline is only next to itself, not divided by the other stories.

The panels offer the chance to explore relative size using relationships among the panel colors and the chance to look for instances of equality in panel size or number. For example, Page 28 has a nine-panel display. Each panel on this page is equal (roughly) in size, making the identification of the fractional amounts that a panel represents to the story line more straightforward than if the panels were all different sizes. There is the opportunity to discuss what fractional amount each color represents on a certain page, and think about how much time (total numerical value) is being dedicated to each story line on a page. For teacher reference, on Page 28, the nine panels can be represented with the following fractions: 1/9 of the story is red, 5/9 of the story is yellow, and 3/9 of the story is blue. Once it is brought to students' attention that we could keep track of how much of a page is dedicated to each story line, this could be something that is continuously recorded, either individually or collectively, over the course of the entire book. Then, students could compare the overall prevalence of story lines throughout by finding the total space dedicated to each story line in the novel as a whole.

To enrich this experience, students can start thinking about the relationship of colors on pages in which the panels are not of equal size. For example, on Page 29, the red panels seem to take up ⅓ of the story and the blue panels ⅔ of the story. However, there are two red panels that compose that ⅓ and they are not the same size. So a teacher can ask students what those two sizes of the red panels might be. The same could be asked of the three blue panels composing ⅔ of the page. Teachers might also ask students to discuss the relationship between the red and blue on the page and how the ⅓ and ⅔ are related. Next, we provide another prompt to continue exploring the relationships of colors to the story lines from the opening section of the book:

- Review back to Page 1 of *The Magic Fish*. This page has all 3 colors, and some students might say that each color could be identified as ⅓, ⅓, ⅓ of the story. However, a teacher can pose the following to embark on other discussion paths: "To me, it looks like blue is ½ of the colors represented, and the other half is shared equally among yellow and red. What fractions might represent yellow and red if blue is ½?" Since yellow and red are equally sized, students might offer that by together representing ½, they each would be ¼.

Section 3 (Pages 41–64): What Would You Call This Fraction?

Often, a simple visual comparison of fractions is all that is needed to solve fractional problems as they are written on state assessments and other diagnostic multiple-choice exams. This section may offer the chance to remind students of this. In other words, not every question involves an algorithm. This section is a great place to practice students' ability to quickly reason about the relative size of fractions and name them in relation to a changing whole. An interesting set of panels occurs on Page 66 where there are three rows of panels; however, each row has individual panels of different sizes. One question surrounding this page might be "What portion of the story on Page 66 does the panel in the bottom left corner (Tiến's parents embracing) take up?" Here students might productively struggle to name the fractional portion because all of the panels are of a different size. In this case, they may reason that they can break up each row into "common chunks" and estimate that the bottom left panel is approximately 1/12 of the story on that page. Continue these non-algorithmic reasoning activities throughout the section, or throughout the remainder of the book, to encourage conversation and sensemaking without the support of formal steps.

Section 4 (Pages 64–85): Exploring How Fractions Change as Wholes Get Bigger and Smaller, and Fractional Relationships

Pages 64 to 65 are a two-page spread that uses all three colors and different sized panels to add to the complexity of how we discuss relationships between story lines in *The Magic Fish*. Teachers can ask students to first look just at Page 64 and analyze the fractional relationships there: *How much of the story does blue represent?* Students might offer roughly ¾, or might reason about the fractions looking at the panels as groups of three vertical columns of ⅓ + ⅓ + ⅓. If they use this vertical approach, they might offer the first group of vertical panels is three blue that make up ⅓ of the story line, the middle set of ⅓ panels is ⅔ blue, so ⅔ of ⅓ or 2/9. However, if the bottom middle column's red panel was swapped for the third column's top right blue panel, it would mean another total column of ⅓ of blue. So ⅓ of the total column groups is blue, plus another ⅓ of the total column groups is blue means ⅔ of the story line on this page is blue. Continue with questioning on just Page 64: *How much does red represent? How much does blue represent compared to red?* Then change the whole to represent both Pages 64 and 65, which also introduces the color yellow. As students think about both pages together as the new whole, discuss how this changes when identifying fractions and the fractional relationships. Use questions such as the following: *What do you first notice when we expand the whole to both pages? Without calculating and just by looking, how do you think the relationship of blue and red might change?*

Additionally in this section, Page 82 uses all of the same color (blue) but different sizes of panels. While the blue means all of the narrative here is within one story line, Nguyen used different sizes of blue panels to convey different emphasis. For this scenario, pose the following questions: *Why might different sizes be used on this page? How could you represent each panel with a fraction? How could you mathematically represent how the fractions (and so, the panels) relate to one another? Identify a fraction each panel represents and how these fractions relate to one another.*

Section 5 (Pages 86–116): Noticing and Wondering about Rational Numbers and Zooming-in Approach Revisited

As students are reading, space should continue to be held for them to articulate what they are noticing and wondering about the numbers and mathematics of the text, and refer back to the initial chart that was established in the "While Reading *The Magic Fish*" section. Teachers must be careful noticers and remain open to letting mathematics discussions follow the students' interests and contributions. One thing students or teachers may draw attention to is Tiến's age. On page 87, we learn that Tiến turns 13 this year. The number 13 serves as a starting point for many discussions about numbers and their properties. For example, 13 is indeed a rational number (students might note it as 13/1), but it has other properties as well. It could be described in the following ways: 13 is prime (only divisible by 1 and itself), 13 is odd, 13 is a whole number, 13 is a counting number, and 13 is a natural number following 12 and preceeding 14. This last fact can be used as a launch for another zooming-in investigation.

In the "Before Reading" investigation, the teacher and students started to explore discussion-based mathematics with multiple ideas, strategies, and multiple accurate answers. Again, the classroom dialogue excerpt from that chapter emphasizes that as the mathematical parameters become more detailed, student responses may also need to be adjusted to reflect the changing parameters. In the Section 1 investigation, students were invited to explore all the numbers between 1 and 2. Now, students can dig into the prompt of finding as many possible rational numbers between 12 and 14. Qualifying the zooming-in with this parameter of finding rational numbers will help teachers to assess if their students are understanding what rational numbers are. After giving students some work time, extend to students questions like the following: *How are the rational numbers between 12 and 13 similar to and/or different from the rational numbers you find between 13 and 14? Is there a method for identifying all the rational numbers between 12 and 14? Is finding all the rational numbers possible?* These questions can lead to a discussion of the task being infinitely long in duration, because the possibilities of rational numbers between 12 and 14 are infinite.

Section 6 (Pages 117–131): Incorporating Identity and Social Themes in Mathematics

This section could extend the open-ended mathematics discussion into discussions of students' identities and social themes impacting their lives that can then be connected to mathematics. The launch for this work could be Pages 120 to 121: *Tiến's mother flies home for the funeral. Her passport/ American citizenship took her eight years to earn, and she sacrificed returning home in those eight years. Was it worth it? Is it fair? What are others in our space going through?*

Incorporating identity is important to the work of funds of knowledge (González et al., 2001), and when we learn of students' identities, we can learn of the social issues that impact their lives. Integrating social themes into a classroom can be tricky for a variety of reasons, including the time it takes to embed social issues with sensitivity and awareness. The work is context dependent and therefore necessitates a keen awareness of the students in the space, their expressed needs and community needs, and the societal issues at the time. The incorporation of identity and social themes into mathematics can eventually lead to social justice mathematics aims and using mathematics to solve relevant problems in students' lives. Our definition of social justice mathematics follows Gutstein's (2003) theme of student empowerment. We attempt to draw on students' mathematical funds of knowledge (González et al., 2001) and prepare as teachers to follow their interests and needs in the classroom in a variety of ways. Our goals are to develop students' mathematical power, foster a positive mathematical self-orientation, and shine a spotlight on their abilities as mathematicians in their own right. We hope that in this mathematically supportive environment, students will be able to identify opportunities in which they can use mathematics and mathematical arguments to improve realities for themselves and others.

Oftentimes we as teachers use the excuse that the content topics do not lend themselves to mathematical opportunities in pursuit of social justice aims. This is not entirely accurate. The mathematical content may be suitable, but what we struggle with is finding the right "stories" to tell. The power of contextualized mathematics stems from the appropriateness of the context. Our failures come not in the identification of the mathematics, but in providing "cheap stories" (Gutstein, 1998, 2003) for our students to explore. Cheap stories are contexts which involve "real-world" mathematics, but either trivialize the context or are only "real" to a small subset of the students that we are teaching in front of. An example of a cheap story might be the various word problem contexts in typical textbooks that involve amusement parks, bake sales, shopping, and so on. While certainly topics from the real-world, they may not be relevant topics to our students' worlds.

The Magic Fish offers a number of potential mathematical contexts that might act as springboards into relevant and important mathematical discussions focused around the valorization of mathematics in the home life of students of all backgrounds, as well as the recognition of potentially transformative mathematics in communities. An example within the text might be the "hidden" mathematical understandings that Hiền must have in order to be able to create Tiên's new jacket for the dance—the one that she cannot finish because she must travel for her mother's funeral. One might also wonder about the distances (in terms of both physical distance and time) that Hiền must travel to return home.

It is difficult to provide instructions in the form of activity plans for an assignment that one might undertake with students in a classroom, because again, the aims of using mathematics to problematize and solve social issues are context-driven. Rather than give scripted offerings that ignore context, we offer guidelines that we have found helpful in identifying and highlighting the mathematics central to the embodied experiences of students:

- Drawing on students' funds of knowledge to learn who students are and learn and connect with their community assets.

- Think mathematically when having casual discussions with students—the mathematics you will want to highlight will likely come out of these conversations or the conversations that students are having with their peers.
- Ask students to identify areas of concern within their communities and be ready to help them solve the problems that are relevant and applicable to them. Teachers may want to ask students if they are inspired by any of the problems that need to be solved in *The Magic Fish*? Why or why not?

Sections 7 through 10 (Pages 132–229):
Reinforcement and Review or Extension

For a reinforcement of previous content, students can apply the investigations from Sections 1 to 5 in the closing Sections 7 to 10. An especially intriguing investigation that could continue throughout the remainder of the book would be the mathematics that was started in Section 2: *compare the overall prevalence of story lines throughout by finding the total space dedicated to each story line in the novel as a whole*. Table 1.2 provided an example of a tracking that could be done by students throughout the novel and into Sections 7 to 10. Another example would be digging into what was started in Section 6 investigation and opening our math space into solving problems that are important to the students and their lived experiences. While this might start with inspirations about the problems that need solving in *The Magic Fish*, it should not be limited to those problems. Rather, students could start to think about what they see in their own lives and communities that might be addressed with mathematics. This will take the teacher listening and helping guide students into the mathematical knowledges and tools they may need to solve problems they see relevant to their lives and communities. This opportunity allows for mathematics to be recognized beyond a subject that happens inside classroom walls, and as applicable to outside, real-world contexts.

For an extension, students could start to explore the overall structure of graphic novels and how they are made. This will be especially important if the teacher decides to have students embark upon the "After Reading: Extending the Graphic Novel" or "After Reading: Writing Our Own Lives" options (see later in this chapter). In these after-reading options, students create graphic novel pages and either extend *The Magic Fish* or focus on a story line from their own lives. Rather than wait until completing the book to begin this work, students could start to work on this project as they continue to read to the end of *The Magic Fish*, and start by examining Nguyen's structure. Students can examine a page in each Sections 7 to 10 and answer: *What is the area of each page? What is the area of each panel on a page? What is the area of the gutter space?* Answering these questions within the structure of *The Magic Fish* will support measurement and geometry skills and then help students to apply this work to the creation of their own graphic novels.

Ongoing Literary Analysis

Students can explore the mathematical concepts above for their impact on the text. For example, teachers can ask students to discuss the concept of infinity and time within the text. *How does time impact Hiền and Tiến, especially in relation to their family in Vietnam?* (Students may answer that Hiền feels guilty for the amount of time that has passed since she saw her family, and Tiến witnesses his mother's guilt while also waiting for the family to save enough money to visit together.) *How does time impact Alera in Tattercoats?* (Alera is trying to work against time, while she tries to enjoy her freedom before the Old Man of the Sea collects on her promise). *How does the use of time create suspense in both narratives?* (Characters are trying to find time, and also stall for time, to tell others their true identities.) These discussions can continue throughout the novel as appropriate, and as more fairy tale narratives are added. Students can take notes on how time impacts the plot, characters, and mood of the narratives (the fairy tales, Tiến's present timeline, and Hiền's flashbacks). *Does it impact one narrative more than another? Why do you think Nguyen chose to use time in these ways?* In working to queer and humanize mathematics, as well as literacy instruction, we suggest teachers let the discussions and student interest drive the end product. Students could create a rough timeline of the novel using textual evidence, showing how the narratives have their own timelines that relate to each other. They might create multimodal presentations on the concept of time in *The Magic Fish* inspired by the zooming in on the number line activity. This analytical product could also be combined with the color analysis described below (and in previous math instruction sections).

Color plays a large role in visual meaning throughout the novel, and teachers can encourage students to notice these moments and analyze them for narrative effect alongside the explorations of fractional relationships. First, teachers should establish that students comprehend how Nguyen is using the colors for different narratives: blue for fairy tales read by Tiến and his family, and red for the present timeline within the novel. In the first section (using our suggested calendar), yellow (used for flashbacks) is shown on the first page, but the scene is of a sky with no indication of time, so teachers can ask students to predict what that color will indicate. Students can discuss the meaning they interpret from the use of color, and justify their answers with textual evidence as they continue reading. This will allow students to analyze what the author is deeming important for readers to notice, and why. These conversations can also include discussions on panel size and shape, so that students are noticing other important visual literacy elements, combining literary and mathematical inquiry, and thinking about the author's craft in multiple ways. Both the use of visual cues and the theme of time should further be discussed in relation to the overall theme of the importance of stories, and how we can change our own stories. Teachers should decide how often they want students to use this handout; we suggest students analyze 1 or 2 pages per assigned section of reading. In the beginning, teachers may want to assign specific pages (as seen in previous math instruction sections and in Table 1.1) while later in the novel teachers may ask students to choose a page per section to analyze that they are interested in. Discussing in small groups and as a class their choices will help reinforce both the mathematical and literacy skills used in this ongoing exercise.

Teachers can ask additional questions when the use of color is more complex: *Why on some of the blue panels is red included* (for example, see p. 12)? Here, students may discuss that the peaches are a different color to draw attention to Alera's connection to her mother who disappeared into the sea. Hiền, who is reading this story with Tiến, is also thinking about her own mother who is across the ocean. This use of color connects the two characters and their sadness from this separation. *When the colors blend together on page 155, what is the author conveying?* The top three panels on this page show the Vietnamese Cinderella story (being told by Hiền's aunt in Vietnam) visually combining with Tiến's story line in the United States by using a gradient of color from dark blue to red. The first panel is blue and shows Tấm being flown through the sky by birds. Her figure breaks out of this panel into the second, which shows a starry sky with power lines on a background gradient from a purple-blue color, going to more red on the right. The third panel is red and shows Tiến's house, and his name in a speech bubble. This further shows how the discussion of mathematics and literary meaning goes hand in hand with this novel. Thinking through how Nguyen created the story takes us to discussing the deeper meaning behind the visual elements that otherwise might be missed. When reading graphic novels it can be easy to gloss over some of the visual elements in order to find out the plot, but by paying close attention greater interpretation can occur.

AFTER READING *THE MAGIC FISH*

Extending the Graphic Novel

As a summative assessment project, ask students to integrate their knowledge of the text features and mathematical skills to create 1–3 additional pages for *The Magic Fish*. This can be completed using paper or by using free online design tools such as Canva. Teachers can provide sample templates with blank panels in different sizes and arrangements, or can provide fractional parameters that the students must adhere to when creating their stories. This activity offers students an opportunity to utilize their knowledge of the visual components of graphic novels to create a new, meaningful narrative. Students can use their graphic novel handout (figure 1.2) for reference, and could also consider how they can use color based on their investigation using Table 1.2.

The mathematical work students engaged in while reading is easily incorporated into the extended narrative and visual activity described above. Through either a presentation or a written artists' statement, students can analyze their own work by considering some of the following questions, and reflecting on how these aspects contribute to the meaning of their piece: *What is the area of each page? What is the area of each panel on a page? What is the area of your gutter space? What is the relationship (or ratio) between your page to one panel; from one panel to another? What is the relationship (or ratio) of filled space to gutter space on a page? How much of a story line is being told on each page and how can each story line be represented as a fraction in relation to one another?*

Writing Our Own Lives

Another option to extend the reading and mathematics investigations is to ask students to create 1–3 pages of a graphic novel of their own life stories. This option directly connects to Page 120 when Nguyen's characters discuss the power of stories. Students can focus on a moment in their life they want to share with others and then see how the mathematics naturally unfolds in their story to make mathematical connections, similar to what was done with the examples in *The Magic Fish*. Or, they can overtly focus on mathematics as the topic of their graphic novel, sharing their own mathematics autobiographies. With this latter topic choice, students could present a mathematics memory(ies) they have enjoyed, or moments that were challenging or felt limiting, or their family's history with mathematics. What emotions does math evoke? Have there been mathematics content areas that have been intimidating and why? Another choice for their own graphic novels could be to extend the work from Section 6 and ask students to share about the problem they hope to solve and how mathematics will help them. In any of these choices, the questions that were posed with the "Extending the Graphic Novel" option should be used.

Summative Assessment through a Mathematics Task

For teachers that do not have the scheduling space to engage in a summative assessment project like the options described above, a final mathematics task for after reading could be used. This task could still provide the opportunity for open-ended response and also showcase specific mathematical content goals. The following after-reading task asks students to show their ability to justify mathematical ideas and reasoning:

- Find 15 fractions represented in *The Magic Fish*. Cite where you found the fractional representations and justify how you know they are fractions. Then, represent your fractions on a number line. Decide on a scale for your number line that best displays all of your fractions. After representing your fractions on a number line, represent your 15 data points in another way.

BEYOND *THE MAGIC FISH*

Exploring Mathematics from Graphic Novel Extensions

One option for an extension activity is to ask students to read and explore the mathematical connections from peers' extension of the graphic novel that were created in the after-reading section, reinforcing and utilizing students' funds of knowledge to explore different mathematics topics that emerge. Students and teachers can decide together what mathematical explorations and problems to investigate, and as with *The Magic Fish,* discuss how those mathematical concepts contribute to narrative meaning.

Exploring Mathematics in Other Texts

Students could also extend the reading work of finding mathematics that emerges that was modeled with *The Magic Fish* and apply it to other narratives, graphic novels, and texts they have been reading. To explore other Asian American experiences in graphic novels, teachers may consider *American Born Chinese* (Yang, 2006), *Skim* (Tamaki & Tamaki, 2008) or *They Called Us Enemy* (Takei et al., 2019). Teachers can pose questions such as: *How do mathematics topics compare among texts? In what ways are the mathematics topics similar in each type of novel? How do mathematical explorations help your literary analysis? How are different Asian American experiences described in each text, through both visual elements and mathematical elements?*

CONCLUSION

The Magic Fish can be just the start to a mathematics journey that welcomes divergent, creative thinking and supports students in seeing the mathematics all around them—in novels, in other classes, in their daily encounters. Although there are moments when we look for the accuracy and precision of one answer to a mathematics problem, if this is the only way mathematics is ever experienced by students, then we are shutting off the subject to many of our learners and limiting their ways to show their mathematical brilliance. We hope that the discussions, problematizing, and creativity that surfaces with the incorporation of *The Magic Fish* encourages students to expect and demand this sort of open-ended experience with mathematics henceforth.

REFERENCES

Ankiel, J. M. (2021). *"Graphic novel/comic terms and concepts" in pictures tell the story: Improving comprehension with Persepolis [lesson plan]*. ReadWriteThink. http://www.readwritethink.org/files/resources/lesson_images/lesson1102/terms.pdf.

Beckmann, S. (2018). *Mathematics for elementary teachers with activities*, (5th ed.) Pearson.

Chapin, S. H., O'Connor, C., & Anderson, N. C. (2013). *Classroom discussions in math: A teacher's guide for using talk moves to support the common core and more, grades K–6*, (3rd ed.) Math Solutions.

González, N., Andrade, R., Civil, M., & Moll, L. (2001). Bridging funds of distributed knowledge: Creating zones of practices in mathematics. *Journal of Education for Students Placed at Risk*, 6(1–2), 115–132.

Gutiérrez, R. (2010). The sociopolitical turn in mathematics education. *Journal for Research in Mathematics Education*, 44(1), 37–68.

Gutstein, E. (1998, April). Lessons from adopting and adapting Mathematics in Context, a standards-based mathematics curriculum, in an urban, Latino, bilingual middle school. Paper presented at the Annual Meeting of the American Educational Research Association, San Diego.

Gutstein, E. (2003). Teaching and learning mathematics for social justice in an urban, Latino school. *Journal for Research in Mathematics Education*, 34(1), 37–73.

Meyer, D. (2019, August 30). Humanizing math class means teaching math like the humanities. *Dy/Dan*. https://blog.mrmeyer.com/2019/humanizing-math-class-means-teaching-math-like-the-humanities/.

National Council of Teachers of Mathematics. (2014). *Principles to actions: Ensuring mathematical success for all*. National Council of Teachers of Mathematics.

Nguyen, T. L. (2020). *The magic fish*. Random House Graphic.

Pavlik, A. (n.d.). *Going graphic: Graphic novels in the language classroom*. https://www.ltu.se/cms_fs/1.178634!/file/Anthony%20Pavlik-Going%20Graphic.pdf.

Pennell, S. M., & Fede, B. (2017). Reading the math on marriage equality: Social justice lessons in middle school. In S. Pennell, A. Boyd, H. Parkhouse, & A.

LaGarry (Eds.), *Possibilities in practice: Social justice teaching in the disciplines* (pp. 93–106). Peter Lang.

Reese, C. (n.d.). *Features of a graphic novel.* https://slideplayer.com/slide/2407749/.

Takei, G., Eisinger, J., & Scott, S. (2020). *They called us enemy–expanded edition.* Top Shelf Productions.

Tamaki, M., & Tamaki, J. (2008). *Skim.* Groundwood Books.

Yang, G. L. (2006). *American born Chinese.* Macmillan.

Yeh, C., & Otis, B. M. (2019). Mathematics for whom: Reframing and humanizing mathematics. *Occasional Paper Series, 41*, 8.

Chapter 2

Take It to *The Limit*

Maximizing Mathematical Literacy and (Re)solving Conflict

Holly Garrett Anthony, Paula Greathouse and Kaylee Gentry

Playing on the lyrics made famous by the Eagles (Meisner, Henley, & Frey, 1975), there is a wealth of opportunities for students to explore mathematical concepts and develop mathematical literacy beyond the boundaries of the classroom while reading *The Limit* (Landon, 2010). By engaging in mental mathematics, exploring the mathematics behind budgets, analyzing the consequences of compounded debt, and immersing themselves within the story line, students' mathematical literacy is both developed and challenged—taking it to its limit.

In mathematics, students solve arithmetic problems and apply mathematical processes in novel contexts. Beyond school mathematics, secondary students will be expected to solve a multiplicity of problems—often those that arise from some form of conflict. In literature, "conflict provides crucial tension in any story and is used to drive the narrative forward. It is often used to reveal a deeper meaning in a narrative while highlighting characters' motivations, values, and weaknesses" (Masterclass, 2021, para. 1). As students read this young adult (YA) novel, having them track conflict offers them an opportunity to analyze this rhetorical device in relation to the movement of the plot as well as other types of problems often familiar to adolescents that get (re)solved.

THE LIMIT BY KRISTEN LANDON

Matt Dunston, 13 years old, has everything going for him. He gets great grades, has great friends, is a whiz at math, gets along with his sisters, and is spoiled by his parents. His life seems perfect, until one day when it is turned upside down when his parents exceed their government-set spending limit. Matt is immediately shipped off to a Federal Debt Rehabilitation Agency to help work off his parents' debt under Federal Debt Ordinance 169, option D. There, Matt finds that the workhouse is not as bad as he thought. In fact, he actually enjoys being there—his room is large, he has his own computer, there is a pool, there is a paddle wall–Wballroom, there are massive televisions, he has his very own workstation, and the other kids are friendly. As Miss Sharlene Smoot (aka Honey Lady) tells him, there is no reason for him to want to leave. However, after hacking into the workhouse's system, Matt learns that not only has his sister been brought to the workhouse as well, but there is a conspiracy between the workhouse and a company that is attempting to reprogram children into super machines—often at the expense of their health and well-being. In an effort to get to the bottom of all of this, Matt embarks on a mission to save his sister, his family, and eventually every kid in the workhouse.

BEFORE READING *THE LIMIT*

Number Talks

As early as pages two to five in *The Limit*, readers find Matt and his friends, Brennan and Lester, engaging in a lively match of mathematical wits as they use mental mathematics strategies to calculate their scores in an unusually scored basketball game:

> "Anyone who gets their hands on the ball can shoot, but we're going to score it different. We all start with . . ." Today was March 12. "Twelve points." A car drove by. It had two eights on its license plate. "Every time you score, you get to multiply by eight. Every time you miss, you have to divide by four." No reason for the four. I just pulled it out of the air. "First guy to a billion wins." (pp. 2–3)

Embedding *Number Talks* (Parrish, 2010; Humphreys & Parker, 2015; Parrish & Dominick, 2016) within the mathematics classroom prior to reading *The Limit* can equip students with the strategies needed to play along with the characters in the text. *Number Talks* are short (five to seven minutes) classroom segments in which students are challenged to solve an arithmetic problem using only mental strategies—traditional algorithms (like carrying/borrowing in addition/subtraction) are discouraged and no pencils or paper are allowed. The teacher posts a problem (e.g., 63 − 37) and asks students to solve the problem mentally to themselves and to provide a discrete hand signal in front of their chest (typically a thumbs up) to indicate they have a solution. Students who arrive at answers quickly are further prompted to find a second way to solve the problem (and signal by adding a pointer finger to their thumbs up) while the teacher monitors all students and provides ample time (usually no more than one minute) for most (an ideal goal is two-thirds of the class) students to arrive at a solution. After the teacher calls "time," they solicit answers from students (without explanations) and list them on the board. The teacher then prompts students to choose one posted answer and offer a defense, or explanation, for how they arrived at that answer. During this time, the teacher writes the equations to represent the solution strategy as described by the student. The teacher's role is to facilitate, but not lead the discussion. They can ask clarifying questions, but no judgment to correctness or incorrectness of the solution is offered. As peers listen and observe the equations, they determine on their own whether or not the presented solution strategy is correct and they have the opportunity to offer a supporting defense/explanation or refuting defense/explanation. As students are called upon one at a time to share their solution strategies, and as the teacher records their equations on the board (or poster paper), the class collectively determines which solutions are correct and which should be "stricken out" as incorrect. Once an answer has been agreed upon, the teacher can choose to review the incorrect solutions and address any misconceptions that emerged during the *Number Talk*.

Alternatively, teachers can opt to give students a follow-up activity—perhaps in the form of a short writing prompt—to explain which solutions were incorrect and why. To see the transcript of a *Number Talk* in action, refer to *Number Talks* on the *Math for Love* blog (2020) or watch a compilation YouTube video: *Sherry Parrish: Number Talks: Building Numerical Reasoning* (2014). As students become accustomed with the format of a *Number Talk*, the time needed to complete this work is lessened. And, as students encounter more varied problems in the *Number Talks* (i.e., addition, subtraction, multiplication, division, operations with decimals and fractions), they develop more efficient and robust mental math strategies. They also develop skills at recognizing when a particular strategy is more/less efficient based on the structure of the problem. For example, when adding $47 + 39$, students might recognize that the *Make Ten Strategy* is very efficient: take 3 from 39, leaving 36; add the 3 to 47 (make a 10) to get 50; then add $50 + 36 = 86$. In contrast, the *Rounding and Compensating Strategy* is effective, but less efficient: round 47 to 50 and round 39 to 40; then add $50 + 40 = 90$, but I added 3 to 47 and I added 1 to 39, so I have added 4 too many, so I must subtract those from 90 to get 86. We recommend embedding *Number Talks* into classroom instruction daily for a few weeks prior to reading the novel to help build students' skills and confidence with mental math strategies prior to reading. Suggested sequences of problems for *Number Talks* are available in Parrish (2010) or Parrish and Dominick (2016).

Introducing Limits and Budgets

As suggested by the title, limits are a key theme in the novel—government-imposed financial limits, limits to what the physical mind is capable of enduring, limits to where kids can or cannot go in the workhouse, and limits to families' abilities to offset their debt, to name a few. Before reading *The Limit*, students should be provided opportunities to encounter the specialized mathematical language often associated with financial limits or budgets. To elicit this, poll students to see how many have created or managed a budget before, or perhaps balanced a checking/savings account. Ask them to share their stories and encourage others to think about scenarios in life in which sticking to a budget is essential. Some possible answers might include staying under your talk/text/data limit in a cell phone plan, saving for a big-ticket item such as car/home, or paying taxes. Ask students to share stories of when they exceeded their limit and the consequences (if any) of doing so. As students talk, make a list of any specialized vocabulary words that occur naturally in their conversation: *bankrupt, in the hole, in debt, over the line, overspent, in the red, in the black, break even,* or *balance point*. After the sharing session, ask students to list these words in their math journals (or notebook) and write their own definitions as inferred from the contextual clues in the conversation they just had. This will be a useful reference for students as they read the novel.

Defining Conflict and Conflict Typologies

The opening lines of a book may be the most important. It is here that a reader's interest and curiosity is piqued and the author provides an entry point into the story. Kristen Landon wastes no time drawing the reader in and setting up the conflict that drives the narrative. The opening line of *The Limit* reads, "AN EIGHTH-GRADE GIRL WAS TAKEN today" (p. 1). More than likely, the first question that will arise is "Why?" Students will come to learn the answer to this question as they read. Prior to this journey, however, it is key that a definition of conflict is established and that teachers acquaint students with the types of conflict they will encounter within the novel.

There are three types of conflict that students can be asked to identify and analyze as they read *The Limit*: character versus character, character versus society (the government), and character versus themselves. In order to have students begin to consider these types of conflict, two introductory approaches are necessary. The first is to ensure that students are all working

Term	Know it Well	Think I Know It	Don't Know It	Definition
Conflict	X			A fight between two people

Team definition: A disagreement between two people or struggle between two things.

Final class definition: A struggle between two opposing forces.

Figure 2.1 Sample Vocabulary Rating Scale Graphic Organizer. *Source*: Created by authors

from the same understanding of the word "conflict." One way to accomplish this is to have students participate in a modified vocabulary rating scale activity. Teachers can provide students with a graphic organizer that asks them to not only rate their understanding of the word "conflict," but draw on their background knowledge to define the term (see figure 2.1). Provide students with a few minutes to individually complete the chart within the graphic organizer. Once all students have rated and defined the term, place students in teams of two or three. Ask each student to share their definition of *conflict* with their teammates. Once each student has shared, ask the teams to collaborate and create a new working definition of *conflict*, combining elements of each team member's individual definition. In order to ensure that all students are contributing to this definition, teachers can provide each student with a different color pencil or pen and ask that they use this as they craft their collaborative team definition. This provides the teacher with a visual of student contribution.

Once each team has developed a working definition of the term "conflict," ask one member from each team to write the definition on the board, or in a place that is visible to all students. For example, a teacher can post several pieces of chart paper (one for each team) throughout the room and ask that each team record their definition on a specific sheet. Once all definitions are recorded, teachers can have students complete a gallery walk, reading other teams' definitions. As they do this, ask students to note similarities and differences between their team definition and other teams' definitions. This gallery walk should take no more than five minutes. At the end of that time, bring students back together as a whole class and have a discussion on the similarities and differences they noted. As students share, the teacher could be charting these on a separate piece of chart paper or on the board. When all similarities and differences have been exhausted, ask students to begin to craft a whole-class working definition. While the students will be driving this definition, it is important to note that teachers should ensure that all necessary elements are included in the final definition. Once a final definition is crafted, ask students to record this definition in their class notebooks or have a student copy the definition on a piece of chart paper posted in the classroom. The goal is to make sure that students have direct access to this definition as they read.

After students have established a working definition of the term "conflict," it is important to ensure that they have an understanding of each type of conflict they will be exploring as they read. Teachers can extend the modified vocabulary rating scale activity and ask students to work through each type of conflict following the same steps outlined earlier: character versus character (two characters struggling with each other), character versus society (character struggling with an element of society), and character versus themselves (character struggling internally).

WHILE READING *THE LIMIT*

Analyzing the Basketball Game

Before reading, students engaged in *Number Talks* to develop their mental math strategies. As they read the novel, the first opportunity to apply the learned strategies is in the scoring of the competitive basketball game between Matt and his friends (pp. 2–5). After reading this section of the text, provide students with the following questions. Mental math should be encouraged, but allow the use of pencil/paper to aid their calculations; calculators should not be permitted:

- Use Matt's scoring system to check his math (p. 4). Is his score of 12,582,912 accurate given the hits/misses he reports?
- What combination of hits and misses would give Brennan a score of 1.5? Are there other combinations? What pattern are you noticing?
- What combination of hits and misses would give Lester a score of 12? Are there other combinations? What pattern are you noticing? How is this pattern similar/dissimilar from the one observed in Brennan's scoring system?
- Matt says that he only needs three more baskets to score one billion (p. 4). How can you use mental math to support/refute his claim?
- Lester says that he's been multiplying by five (p. 5), and that according to his scoring system, Matt needs six more baskets to win. How can you use math to support/refute his claim?

- Brennan says he's been multiplying by 3.5, and that Matt needs 9 more baskets to win (p. 5). How can you use math to support/refute his claim? How can you use your work for Lester's scoring system from Question 5 to quickly determine if Brennan's claim is accurate?

Students should conclude that the mathematics in the novel is accurate and affirm with their own calculations that "even during a basketball game Brennan and Lester processed more numbers through their brains than air through their lungs. I [Matt] probably did too" (p. 2).

Averages, $d = rt$, and Integers—Oh My!

As students follow the not-so-yellow-brick road throughout the novel—per your predetermined reading schedule—additional opportunities to explore mathematical concepts will present themselves. One such opportunity occurs in Chapter 5 when Matt wriggles out of Honey Lady's grasp in the backseat of the limousine and turns back to the window to begin "calculating the average number of windows for the buildings we passed" (p. 47). Students can be tasked with an interesting (yet simple) mathematical exercise of first counting all of the buildings/homes they pass on the way to school from home each day, and then counting the number of windows on all of those buildings/homes, and finally calculating the average number of windows for the buildings they pass each day. A modified round-robin version of show-and-tell (two minutes maximum per student) can be facilitated as a whole-class sharing activity for students to showcase their work with peers.

Another opportunity to engage in mathematics within the novel arises in Chapter 6 when Matt contemplates an escape through the lobby of the workhouse (pp. 56–57). Matt recalls "distance equals rate times time—one of the most basic math formulas in the world" (p. 56) and after estimating the distance and his running rate, he determines he will need about 10 seconds to cross the lobby and reach the door: "Dividing the distance by the rate gave me a time of 3.47 seconds. I figured I should round up to five or six seconds to allow for the fact that I'd have to turn a corner and dodge a chair or two" (pp. 56–57). Teachers can leverage this to help student pairs calculate their own "escapes" from a location within their school. For example, several post-it notes (one for each student pair in the class) can be posted on the floor of the gym (or hallway). Some other location should be designated as the "escape door" that students are working to reach. Students must locate their designated post-it note and then estimate the distance from their location to the escape door. They would then be asked to measure the actual distance and record it on their note. Finally, students will calculate their running rate (in feet per second) by timing each other as they run a designated (known) distance—for example, from half court of a basketball court to the end line is 47 feet. Ask student pairs to use the data for each of them to calculate their average running rate. Using the distance from their post-it note to the escape door, and the average running rate for their team, students will use the formula $d = rt$ to calculate how long it will take their team to escape. Once all pairs have completed the activity, facilitate a whole-class discussion and ask the teams to order themselves according to which pairs would escape first, second, third, and so on, based on the times they calculated. Variations of this activity can be done in which (1) teams move from post-it to post-it and calculate all of the escape times or (2) time and distance are provided and they calculate how fast they would need to run to escape and they then time themselves to see if they can achieve that rate.

One final opportunity occurs in Chapter 18 when Matt ruminates, "The chances of my entire family ever living at home together again were about as good as the product of two positive numbers coming out negative" (p. 184). This offers a segue into a discussion of operations with integers and why the product of two positive numbers is positive and the product of two negative numbers is also positive (not negative). Students have often memorized integer "rules" without understanding the concepts. Why is a "negative times a negative a positive"? Before answering this question, specifically, we suggest a mathematical investigation into all of the product rules for integers. Begin by listing the product rules:

- A positive times a positive is a positive. Example: $3 \times 3 = 9$
- A positive times a negative is a negative. Example: $3 \times -3 = -9$
- A negative times a positive is a negative. Example: $-3 \times 3 = -9$
- A negative times a negative is positive. Example: $-3 \times -3 = 9$.

Using what students already know about the meaning of multiplication—thinking of 33 as "three groups of three" or "three people each having $3"—it is easy to explain why a positive times a positive is positive since they would have $9 altogether. Likewise, when thinking of why a positive times a negative is negative, we can conceptualize this as "three people each owing $3" so their total combined debt is $9 as represented by –9. Using the commutative property of multiplication—$ab = ba$—and the discovery we just made that a positive times a negative is negative, we can conclude that a negative times a positive is also negative. Lastly, we need to address the interesting case in which a negative times a negative is positive. There are a number of mathematical proofs (and the distributive property) that could be used here, but we recommend asking students to work with patterns to make this discovery. List a series of equations and ask students to observe the pattern:

$-3 \times 3 = -9$
$-3 \times 2 = -6$
$-3 \times 1 = -3$
$-3 \times 0 = 0$
$-3 \times -1 = ?$

As students consider the answer to the last equation posed, help them explore the pattern among the previous products: −9, −6, −3, 0, __. Ask, "What follows in this pattern?" Students should conclude that positive 3 would be next in the sequence and is therefore the product of −3 and −1. We see, then, that a negative times a negative is a positive. Ask students to continue the series of equations presented to explore:

$-3 \times -2 = ?$
$-3 \times -3 = ?$

As a concluding activity, ask students to explain to a peer why a negative times a negative is a positive. This can be done through a simple turn-and-talk.

Tracking Conflict Typologies

As discussed in the before-reading section, the opening line of *The Limit* immediately introduces the reader to conflict. However, this statement alone does not indicate the typology of conflict that led to the girl being taken, nor does it identify the only type of conflict that readers will encounter as they read. Because the story presented in this YA novel is framed around and driven by conflict, it is important that students not only track conflict typologies from the onset of their reading, but identify how the conflict is resolved, if at all, and the ways in which conflict drives the narrative. One way to accomplish this is through the use of a tracking graphic organizer. Teachers could approach this in one of two ways. One way is to ask students to individually track each conflict typology as they read. Alternatively, teachers could create student teams and ask each team to focus on one conflict typology. Figures 2.2 and 2.3 offer a visual of what this tracking graphic organizer might look like in terms of format and completion after reading the first two chapters for both suggested approaches.

As students continue to read, ask them to add to their tracking graphic organizer. Some suggested examples of each conflict typology presented in the narrative that teachers could guide students through include the following:

Character versus Character
- Mom versus Honey Lady (Miss Smoot)—when she comes to take Matt (35–39)

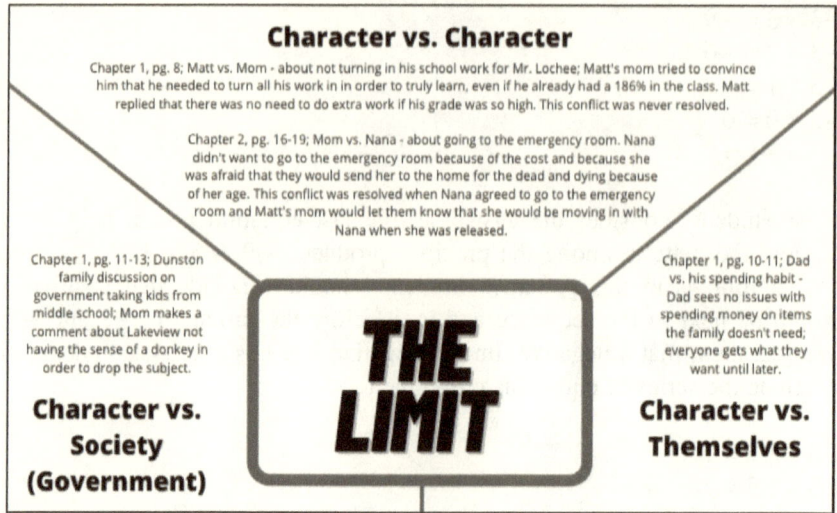

Figure 2.2 Sample Tracking Graphic Organizer for All Conflict Typologies. *Source:* Created by authors

Character vs. Character

Chapter & Page Number	Brief Description of Conflict	Resolution
Chapter 1, pg. 8	Matt vs. Mom - about not turning in his school work for Mr. Lochee	Matt's mom tried to convince him that he needed to all his work in order to truly learn, even if he already had a 186% in the class. Matt replied that there was no need to do extra work if his grade was so high. This conflict was never resolved.
Chapter 2, pg. 16-19	Mom vs. Nana - about going to the emergency room.	Nana didn't want to go to the emergency room because of the cost and because she was afraid that they would send her to the home for the dead and dying because of her age. This conflict was resolved when Nana agreed to go to the emergency room and Matt's mom would let them know that Nana would be moving in with them when she was released.

Figure 2.3 Sample Tracking Graphic Organizer for One Conflict Typology. *Source:* Created by authors

- Matt versus Gorilla Man—when being taken to fulfill FDO 196-D (38–40)
- Matt and Coop versus Kia, Isaac, and Madeline—using the gym (102–105)
- Matt versus Gorilla Man—trying to get to Lauren (134–140)
- Matt versus Honey Lady (Miss Smoot)—being locked on Top Floor and seeing Lauren (142–145)
- Matt versus Mom—about the family money situation (171–182)
- Matt versus Reginald when they first meet face to face (201–204)
- Matt versus Paige about parents caring (211–215)
- Matt versus Gorilla Man when subdued after breaking pinball machine (242)
- Matt versus Honey Lady when confronting about hacked findings (243–255)
- The Top Floor kids versus Crab Woman when being held captive (260–274)
- Everyone versus Miss Smoot when they realize what she has done to them (277–282)

Character versus Society (Government)

- Mom versus Lakeview society—comment about them not having the sense of a donkey (13)
- Dunston Family versus the Limit—went over (23–26)
- Matt versus Government—he gives Jessica the flash drive and directs her to share with the media to expose the happenings in the workhouse (252–255)
- Honey Lady (Miss Smoot) versus Government—she tries to escape government capture (276–282)

Character versus Themselves

- Dad and his spending habits (10–11)
- Matt's struggle in deciding to leave the workhouse (58)
- Matt's internal struggle with express shipping (112–113)
- Matt's struggle with deciding to hack into the workhouse computer system (123–125)
- Matt's, Coop's, and Jeffery's struggles with deciding to go outside (159–166)
- Matt's struggle with his decision to spend money out of frustration with parents (pg. 185–190)

As students read and complete their tracking graphic organizer, it is important that they have an opportunity to discuss what they record. Teachers could ask students to share their notations with a partner or in small teams at the end of each chapter, or after a chunk of chapters have been read. Or, teachers could have a whole-class discussion about what students noted.

While many students may have the same information recorded on their tracking graphic organizer, some students may have missed some of the examples from the text. In an effort to ensure that students are identifying the conflict typologies, teachers could bring students together as a whole class and ask them to share their notations. As they share, the teacher could direct students to add any notations discussed that they did not originally include on their tracking graphic organizer.

AFTER READING *THE LIMIT*

Simulating *The Limit*

After reading the novel, engage students in a simulation of *The Limit* to further their understanding of (imposed) budgets, debits/credits, and the mathematical difficulty (impossibility) of working your way out of debt that is compounding at a rate greater than it can be repaid/forgiven. Have students revisit Chapter 17 in the novel (pp. 170–181) to review the conversation between Matt and his mom when he escaped from the workhouse to inform his parents of his sister's seizure. In that conversation, Matt inquires about the status of the family's debt and learns that his performance at the workhouse has increased his family's monthly allowance, but their debt has also grown. Matt asked, "Why didn't they send me home as soon as you got the new limit?" to which Mom replied, "They couldn't. . . . If you came home, we could drop back down to the old limit. *That's* the one we have to get under in order for you to be able to come home" (pp. 174–175) (emphasis in original). This excerpt opens the door for an in-depth mathematical look at the complications of removing debt that is accumulating faster than you can repay it. For this two-week classroom simulation, assign students to "families" of four and assign a budget limit that must not be exceeded over the course of the week (e.g., $100). Work collaboratively with the class to establish a list of classroom behaviors or practices that yield income (credit) or incur cost (debit). Some examples might be (1) each time a member of the "family" asks the teacher to repeat directions already given to the class, they incur a debt of $20, or (2) each time a family member volunteers to answer a classroom question (and is called on by the teacher), the team yields an income of $10. We recommend that the list be limited (pun intended) to behaviors or practices and not related to performance or grading. For example, the family should earn $10 if a family member is called on to answer a question (after volunteering) even if the answer to the question is incorrect. Each member of the family will track their individual income/debt in a checkbook register (template is easily located online) over the course of a week with the teacher assisting by helping flag debits/credits that should be recorded. During the week, family members are discouraged from checking with each other to track the status of their spending/depositing as was the practice of the characters in the novel: Matt asked, "Don't you check it regularly? Mom! How can you know how far over the limit we are if you don't check the account? Did you ever check it before we went over?" Mom looked like she'd just been stung by a bee. "I don't like to worry about money. Dad always assures me we have plenty coming in and that he'll take care of everything" (pp. 172–173). At the end of the week, the family members will meet in groups to compile and calculate their family's balance. Determine how many families went over the limit and then address ways they can work their way out of debt in the following week. As families strategize ways to work out of debt in the next week using the same debit/credit rules already established, introduce a new element for the second week that involves *Life Happens* cards that must be drawn once each day of the following week for each family (not for each family member). The teacher creates the *Life Happens* cards that mimic some of the circumstances in the book. Examples might include car breaks down and $50 repairs are made, Mom earns $30 cash in marketing business, Mom spends $40 on inventory for business, Dad loses $25 contract, Dad plays golf for $10, Dad earns $20 commission on sale, Baby Sister needs new $15 toy, and similar. Create enough cards so that each family draws five cards the following week (one for each day of the week); so if you have 6 teams in your class, you should create 30 different cards. Depending on whether students have previously studied simple/compound interest, the teacher can choose to include cards that include situations requiring the calculation of interest rates. The second week unfolds in the same way as the first week with the addition of drawing a *Life Happens* card each day. At the end of the second week, families should reconvene to see if they successfully worked their way out of debt. Facilitate a whole-class discussion of the results to highlight which teams stayed (or moved) below the limit, which teams went over the limit (and by how much), and discuss the impact of adding the *Life Happens* cards to the simulation. Team members should individually submit a Budget Report as a culminating assignment that includes the compiled family checkbook register with mathematical calculations. The Budget Report should be accompanied by a reflective paper in which students respond to the following two prompts excerpted from the novel:

1. Matt claims, "Managing the account isn't calculus. Heck, it isn't even algebra. Basic, simple arithmetic," to which Mom replies, "Oh, no. It's much more complicated than that. It's got all that interest to calculate and the fines and fees to figure out" (p. 173). Reflect on your experience in the simulation to determine whether you agree with Matt or Mom. Use mathematics to support your decision.

2. Matt bemoans, "The dollars in our account disappeared faster than I could think of them" (p. 181). Relate this comment to your family's experience in the past two weeks. How did Matt's comment play out (or not) for your family?

Show Me What You Read

As students read the novel, they tracked the different conflict typologies manifested within the story. For this after-reading activity, students are asked to move beyond this identification and into an analysis of the ways each typology of conflict influenced the narrative. One way that students could demonstrate this critical thinking is through the creation of a One-Pager. This visual approach offers students an opportunity to be creative as they show what they read.

One-Pagers ask students to create a visual representation of the text in a way that highlights a specific literary element, or elements, that were explored through their reading. Although a teacher could locate several versions of One-Pager directions with a simple search on the web, there is no prescriptive set of directions in the construction of a One-Pager; teachers will establish different purposes for reading which will guide what is included. For this activity, teachers could ask students to include the following on their One-Pager as a means to ensure representation of conflict and its influence on the plot of *The Limit* (see figure 2.4 for example):

- the title of the book;
- the names of the conflict typologies and two to three words or phrases that describe each typology;
- at least one image per conflict typology;
- a quote that shows each conflict typology with a sentence or two explaining how that conflict influenced the narrative.

In the during-reading section, two approaches to tracking conflict were suggested: students track one conflict typology throughout their reading or students track all conflict typologies as they read. If the former tracking approach was utilized, teachers could modify the elements presented earlier. For example, if students each tracked a single conflict typology, then that typology would be the sole focus of the One-Pager.

Once students have finished crafting their One-Pagers, have them share their final product with their peers. Teachers can approach this sharing in several ways. For example, the teacher can place the One-Pagers around the classroom and students could do a gallery walk. Another approach could be having students present their One-Pagers in small teams before placing them around the room. Or, the teacher could ask students to create a video presentation, sharing their One-Pager and describing its elements. The teacher could then play the videos during class as a way to share students' products.

Figure 2.4 Sample One-Pager. *Source*: Created by authors

BEYOND *THE LIMIT*

Financial Literacy

The root cause of conflict in this YA novel is the monetary limit that the government has placed on its citizens. As an extension activity, teachers could help develop students' financial literacy through several approaches.

Living within a Limit

Like Matt's family, all families in *The Limit* must be financially literate in order to obey the government's order. For this activity, teachers can ask students to consider whether or not they could live within a monthly monetary limit. Begin by asking students to choose a career that interests them and have them locate the average yearly salary for an entry-level position in that career. Sites like CareerOneStop.org would be a great place for students to begin as this site offers the opportunity to input not only your desired career but the location of where you would like to work. Once students have determined what their monthly salary would be, ask them to calculate the taxes that would be deducted from this amount. Next, ask them to consider how much insurance would cost and require them to deduct both the tax and insurance amounts from the monthly salary. Students may get excited at first thinking that their take-home pay is pretty good and one that they could easily live on. Remind students that while the amount calculated is their monthly net pay, they must also consider other expenses. Have them put together a monthly grocery list; estimated cost for monthly car payment and gas for their vehicle; housing costs such as rent or mortgages; estimated cost of monthly utilities such as electric, water, garbage/sewer, gas, cable television, internet, and so on; and any costs associated with entertainment or clothing expenses. Ask students to subtract these monthly expenses from their net pay to determine if they could actually live within their set limit. In teams, or as a whole class, students could share their findings and a discussion around financial literacy could ensue. From there, teachers could guide students through a discussion on *wants* versus *needs* and invite a local financial planner to visit and share strategies with students on how to be financially responsible.

As an extension to this activity, teachers could ask students to create a "Limit Confessional" video. In this video, students could share their insights into this activity and describe ways in which they could be successful in staying under their limit. Through this confessional, students will have an opportunity to reflect not only on the activity itself, but also on what it means to be financially literate.

Revisiting the Text

Another way that teachers can help students develop financial literacy is to revisit the book and discuss instances in which finances impact the plot. Students could be teamed for this activity and the text could be jigsawed. Below are some examples from the text and some guiding questions that teachers can pose to students:

- Chapters 1 to 7: Describe what happens after a family goes over their limit (1, 25–40, 44–45, 47–53, 64–69).
- Chapters 8 to 14: What is the relationship between money and those living on the top floor (109–113)?
- Chapters 15 to 21: Describe Matt's parents' financial plan for reducing their debt. (170–182). Describe how Matt reacts to his mom's financial plan (pyramid marketing scheme) and his reasoning for acting this way (p. 185).
- Chapters 22 to 26: Matt develops a plan to help solve his family's financial troubles. Describe this plan (229–255).

CONCLUSION

As students read *The Limit*, they were asked to solve mathematical equations using numbers found within the text and to identify and analyze conflicts that manifested and drove the narrative. By exploring limits—mathematically and metaphorically—students further developed their mathematical literacy. They saw concrete, practical applications of mathematics (mental math, budgets, patterns) as well as abstract applications of mathematical thinking (logic, reasoning, and problem-solving) that they can use to understand the characters and their lives. The mathematical tasks and the analysis of the literary element of conflict presented in this chapter asked students to identify and analyze problems. These activities encourage students to take their thinking to the limit.

REFERENCES

Humphreys, C., & Parker, R. (2015). *Making number talks matter: Developing mathematical practices and deepening understanding (Grades 4–10)*. Stenhouse.

Landon, K. (2010). *The limit*. Aladdin.

Masterclass. (2021). *What is conflict in literature? 6 different types of literary conflict and how to create conflict in writing*. https://www.masterclass.com/articles/what-is-conflict-in-literature-6-different-types-of-literary-conflict-and-how-to-create-conflict-in-writing

Meisner, R., Henley, D., & Frey, G. (1975). Take it to the limit [Recorded by Eagles]. *On One of These Nights* [Album]. Asylum.

Parrish, S., & Dominick, A. (2016). *Number talks: Fractions, decimals, and percentages*. Math Solutions.

Parrish, S. (2010). *Number talks: Helping children build mental math and computation strategies (Grades K–5)*. Math Solutions.

Chapter 3

Extracting Mathematical Topics Embedded in *Holes* and Examining the Text through a Critical Lens

Marilyn E. Strutchens, Mike P. Cook
and Brea C. Ratliff

Holes (Sachar, 2000) is filled with rich opportunities for a mathematics teacher to engage students in reasoning and sensemaking related to probability, statistics, proportional reasoning, measurement, and other mathematical topics. Throughout the story the main character, his family members, peers, and others encounter situations where mathematics is alluded to, but not made an explicit part of the story. For example, Stanley Yelnats and the other boys at Camp Green Lake dig holes all day that are five-feet wide and five-feet deep. The provision of the dimensions of the holes gives teachers an opportunity to create problems related to the volume of a cylinder, rate of the time it takes to dig a hole, and others.

The major plot of the story focuses on Stanley Yelnats, who is wrongfully convicted of a crime and sentenced to a juvenile detention facility, in which the detainees are largely children of poverty and who do not have advocates for their well-being. This plot justifies a need to examine the text through a critical literacy lens and employ social justice pedagogy. Thus, *Holes* is the perfect medium for enabling students to develop two types of agency: (1) mathematical problem-solving agency, which entails students making decisions about their participation in mathematics, making sense of problems and persevering to solve them, reasoning abstractly and quantitatively, constructing viable arguments, and critiquing the reasoning of others (National Council of Teachers of Mathematics (NCTM), 2015); and (2) sociopolitical agency, in which students use mathematics as a tool to analyze seemingly unjust situations and take actions accordingly (Gutstein, 2006; Aguirre et al., 2013; Strutchens, 2018).

A CRITICAL LITERACY OF READING AND MATHEMATICS

For the purposes of the activities in this chapter, we conceptualize critical literacy as a way of viewing and interacting with texts and as a set of skills for analyzing texts. We draw on critical literacy as a way to help students interrogate the interrelationship between, and the inseparability of, literacy and power. Through this critical literacy lens, readers are encouraged to question and challenge power relationships and the ways in which power dynamics are constructed in, and on, the world (Harste, 2003; Janks, 2010; Luke, 2021).

Drawing on a critical literacy lens, students can engage in sustained analyses of a variety of power dynamics which can include the systems and structures that surround them, and the issues of power represented in literature (e.g., recognizing and questioning multiple meanings, the positioning of authority, the prescription of dominant discourses, etc.) (Behrman, 2006). In these ways, young adult (YA) literature, such as *Holes*, can provide students with "settings of experience" (Sams & Cook, 2019) in which to interact with the representations of the world, to develop critical skills and dispositions, and to apply those to challenge instances of inequity and power imbalance surrounding them (Boyd & Darragh, 2019).

Given the variety of mathematical concepts embedded within the text, *Holes* is an excellent novel through which teachers can help YAs to see how developing facility with mathematics can expand their professional opportunities, help them to understand and critique the world, and experience the joy, wonder, and beauty of mathematics itself (NCTM, 2018).

HOLES BY LOUIS SACHAR

Holes is the story of three interconnected narratives spanning time, generations, and geography but centers around the protagonist, Stanley Yelnats, who is wrongfully convicted of stealing a pair of shoes and subsequently sentenced to Camp Green Lake, a dried-out and sun-scorched lake bed that serves as the setting for a juvenile detention center in Texas. Stanley's family lives in poverty and believes themselves to be unlucky. They blame everything bad in their lives on one ancestor, Elya Yelnats, who emigrated from his home, bringing the family bloodline to the United States.

The second narrative tells the story of Elya, how and why he came to immigrate to the United States, and the curse placed upon him. It is on this curse that Stanley and his family blame their own misfortune.

The third narrative is set a century before Stanley's sentencing, when Green Lake was a lush oasis and supported a small but thriving town. This plotline follows Kate, a white school teacher, and Sam, an African American, who is ultimately killed for falling in love with and kissing Kate.

BEFORE READING *HOLES*

Prior to reading *Holes*, it can be useful to create opportunities for students to reflect on what they already know and to engage in anticipatory ways of thinking. These pre-reading activities provide teachers with insight into what students already know and can do with regard to the thematic and conceptual foci of the text. Because the goal is, in part, for students to develop and apply critical literacy skills to the text—and ultimately to their lives—and to use mathematical thinking as an avenue to meet that goal, we begin with two pre-reading activities.

Fostering Anticipation through Paratext

In the first, students examine the novel's paratext—the book cover in this case—as a way to foster anticipation. Turning the book over in their hands, students' attention is drawn to the book's title and cover image. Here (in the original book cover, although the cover used after the success of the Disney movie adaptation could work as well), readers see a desolate landscape full of holes, one of which has a shovel leaning out of it, a dark star-filled sky, and an abstract representation of a person's head with a cap on backward. After providing students with two to three minutes to examine the images and to consider their meaning, ask students to respond in their reading journals to two prompts: Given the title and the cover image, (1) "What theme(s) do you think this book will cover?" and (2) "What mathematical concepts might you expect to encounter?"

Students may note that the presence of a shovel suggests that someone dug the holes. They may also venture to guess that the person's head we see belongs to a teenager, whom they might assume has dug the holes. Such a line of thinking could lead to questions such as the following: *Why would teens dig holes? Is it of their own choosing or because of someone telling them to?* Students may also draw on the landscape and consider why anyone would be in such an unforgiving place, simply to dig holes. While students may not anticipate the plot of the novel, it is important for them to ask questions and use what information they have to interpret, interrogate, dig deeper, and make inferences. These skills lend themselves well to critical literacy and to mathematical ways of thinking.

In addition to completing the paratext activity, students need time to consider the important themes and concepts within the novel, especially those they will be expected to engage with throughout the unit. While there is no shortage of methods for introducing students to such information, one way this can be accomplished is through pre-reading stations. To prepare for reading *Holes*, teachers can divide the class into six groups, with each group assigned to begin at one of the stations. Station 1 includes a book trailer and book reviews. Station 2 includes an article and statistics on poverty and socioeconomic stratification. At Station 3, students are provided definitions and contemporary examples of terms such as equity, oppression, and racial disparity. Station 4 introduces students to the critical lens of power—or the critical orientation to the power dynamics represented within texts and those that surround the text's creation—they will be asked to apply moving forward. Station 4 also includes a review activity on proportional reasoning. And at Station 5, students are provided with helpful information on probability and statistics. Depending on the teacher's goal, groups can spend anywhere from 5 to 20 minutes at each station before rotating until they have visited all 5 stations. After completing all stations, each group can take five minutes to debrief and reflect upon questions such as the following: *In what ways has reviewing and learning about this information positioned you well to engage with the novel and the themes and concepts you'll encounter in this unit?*

Luck versus Probability

For this pre-reading activity, shift students' thinking slightly from concrete components of the text itself to a mathematical concept that is central to the plot—luck versus probability. As noted previously, Stanley Yelnats's family believes themselves to be cursed and that nothing they can do will change their lot in life. Rather than considering the myriad of factors working against them (or the probability that negative, rather than positive, things will continue to happen to them), they resign their fate to luck, to the bad fortune of having a "no good, dirty, rotten, pig-stealing, great great grandfather" (p. 7). In order for students to question the factors at play in the Yelnats's (mis)fortune and to interrogate the issues of power embedded within the novel, students could consider luck versus probability through both a mathematical lens and through the lens of linguistic power (e.g., subscribing to bad luck benefits those in power, where acknowledging probability leads to questioning and working to interrupt inequity).

In this activity, the teacher can pose a series of increasingly focused and relevant questions to students and provide sufficient time for them to consider each and respond to them in their journal. Begin by asking, "Are the things that happen to us based solely on luck, or are there forces that contribute to the probability of good and bad things happening in our lives?" After students have had a few minutes to think and to respond in their journals, ask, "Why might it be important to consider the difference(s) between luck and probability? Why might that matter to individuals and to the masses?" Again, once students have had adequate time to think and to write through their responses, turn their attention to specific systemic inequities that are often positioned as "luck" by some and as negatively impacting "probability" by others.

In preparation for this activity, teachers can guide students through reading and discussion, both whole class and small group, of a variety of popular press articles and published statistics (e.g., *New York Times*, Bureau of Justice Statistics, National Archive of Criminal Justice Statistics, *Smithsonian*, Economic Policy Institute, and so forth) on income disparity, the ongoing impacts of racist policies on employment and education, and the stark demographic imbalances in juvenile and adult criminal justice programs and outcomes. Similarly, this provides a useful opportunity for teachers to help students develop and transfer critical analysis skills between texts, and between texts and the world. Teachers might pose the following: *Taking into account what you already know about capitalism and the distribution of wealth and resources, or about criminal justice statistics, list a few examples of the variables that influence our "luck," that is, the probability that we will experience good or bad outcomes. Why did you select these variables or forces, and how do you think they impact our lives?* After grappling with these critical social issues together, students can point to racial, ethnic, and gender identities as influencing outcome probabilities. For example, students may draw on data suggesting racial segregation of neighborhoods and property values. Likewise, they may cite data arguing that white students graduate at significantly higher rates than black students. They may also note that African Americans are almost six times more likely to be incarcerated than whites.

The overall goals of the anticipatory set are to activate student prior knowledge, to begin focusing their attention on important themes and concepts featured in the novel, and to assess where they are currently with regard to readiness for planned activities. After completing what students may initially view as two disparate sets of activities, we pose two final questions to students as a way to help them see the connections:

> Now that you've had a chance to think about probability and luck, specifically through manifestations of power imbalances, return to your initial thoughts on the novel's cover image. *What themes do you believe will be covered in this novel? What mathematical concepts do you believe we'll discuss?*

We ask students to compose a less-formal reflective essay, where they engage in this reflective synthesizing individually, rather than in groups, as it (1) offers each of them the opportunity to create threads for themselves between the activities they have completed and those in which they will soon engage and (2) provides teachers a pointed formative assessment before beginning the novel.

More Statistical Reasoning and Sensemaking

In *Holes*, Mr. Pendanski and the warden had a conversation about whether anyone would miss Zero (p. 144). They noted that Zero was a ward of the state; he was living on the streets when they arrested him; he had nobody; and he was nobody. After this discussion, their goal was to erase Zero from the system. Even though this passage occurs late in the text, a discussion about childhood poverty would connect well with the discussion about probability and luck. Moreover, discussing childhood poverty can help students to empathize with Zero later in the story as well as some of the other campers. The State of America's Children (2021) summarized the status of America's children in 11 areas: child population, child poverty, income and wealth inequality, housing and homelessness, child hunger and nutrition, child health, early childhood, education, child welfare, youth justice, and gun violence. These statistics can be used as a resource for students to use data analysis to interrogate the disproportionate number of black and other students of poverty in detention centers. The State of America's Children includes key findings as well as data tables, which are useful for comparing different states. Using statistical reasoning to analyze the data can lead to students developing a deeper understanding of issues related to why some children end up in detention centers. A skilled facilitator of the discussion will help the students think about the myriad of factors that impact students' lives on a day-to-day basis. Table 8: Rental Housing Affordability, FY2020 from *The State of America's Children 2021* is a good table to help guide students' discussion since it shows some of the variables that lead to homelessness, like parents earning minimum wages that do not lead to budgets that can support affordable housing. Teachers can ask students questions related to minimum wages and housing rates. Teachers may also ask questions about wages needed in order to afford a particular house in a particular state.

WHILE READING *HOLES*

Identifying Representations of Power Dynamics/Imbalance

Because the purpose of while-reading activities is to help students make deeper, in-the-moment connections to the text and the mathematics concepts, and to apply those connections to their own lives, one way this can be accomplished is through the use of dialogic journals. This strategy can support students' critical analyses of *Holes* through a lens of power as it can be a place for students to gather text and ideas that they believe are important as they articulate their evolutions as critically literate beings. Students will ultimately return to their dialogic journals as part of their post-reading thematic analysis activity (described in Table 3.1). Table 3.1 provides an example of the dialogic journal, including sample student entries.

The dialogic journal typically only has two columns, but for this example there are three. In the left column, students can collect textual evidence from

Table 3.1 Dialogic Journal Excerpt

Excerpts from *Holes*	How does this excerpt represent or depict power dynamics in the novel?	Connections to power dynamics in your life/world
"Rattlesnakes would be a lot more dangerous if they didn't have the rattle" (p. 93)	In the novel, Stanley doesn't know who he can trust—adults, his peers, himself. There is a constant tension that there are forces at play, those impacting Stanley's life, that he is unaware of.	This reminds me of the idea that the enemy you can see is better than the enemy you can't. We can protect ourselves, to a point at least, against what we know is there. We're completely vulnerable otherwise.
"I'm not stupid. I know everybody thinks I am. I just don't like answering their questions" (p. 99)	In this excerpt, Zero finally opens up and talks to Stanley. He acknowledges that his entire identity at Camp Green Lake has been created for him by others, specifically by the "camp counselors."	We see this in school every day. Teachers, tests, and schools label students, and too often it's simply easier to go along with it than to fight back against a system designed to beat you.
"But perhaps that was part of the curse as well. If Stanley and his father weren't always hopeful, then it wouldn't hurt so much every time their hopes were crushed" (p. 9)	It's like they don't want to stop and ask, "wait, why does nothing work out for us, no matter how hard we work?" It's like focusing on the curse prevents them from truly questioning what's really holding them back.	In the U.S., we find ways to make people feel like their misfortunes are their own fault. The system doesn't want anyone to peel back the curtain. If we can make them feel like poverty and oppression are simply curses or bad luck, we can shift attention away from harmful systems.
"If you take a bad boy and make him dig a hole every day in the hot sun, it will turn him into a good boy" (p. 5)	The Warden and the two counselors use this line as a way to justify the free and unethical (and probably illegal) labor they get from these boys.	I feel like this is representative of the ways adults construct young people for their own benefit. They act as if they know better and expect there to be no questioning their problematic logic.

the novel. In the center column, students respond to each excerpt by describing how and why each selected excerpt represents power dynamics/imbalance in the novel. And, in the right column, students make text-to-world connections and explain the ways in which this, or similar, power imbalances relate to our world.

The excerpts included in Table 3.1 are not comprehensive of the examples of power dynamics and power imbalance within the novel but are instead intended to provide readers with the types of entries students can find. Additional examples include Pendanski making Stanley believe that all his problems are his own fault (pp. 57–58), the idea that manual labor—digging holes in this case—builds character (e.g., p. 27), the comparison between Zero digging Stanley's holes and slavery (p. 132), ongoing reminders that Camp Green Lake "isn't a Girl Scout camp" (p. 14), the adults at Camp Green Lake attempting to delete all mention of Hector Zeroni (Zero) from state records (p. 144), and so on.

Mathematics Opportunities

There are ample opportunities to engage students in mathematical thinking and problem-solving throughout the text, both of which connect in important ways to critical literacy. For example, mathematical concepts and skills including ratios and proportions, statistics and probability, geometric measurement, and others can be explored. Discussing these concepts within the context of the novel can help students see that mathematics is a part of every aspect of life. In addition, providing students with high cognitive demand, multiple entry-level, and relevant tasks enables them to develop problem-solving agency and positive mathematics identities. Student mathematics identities include beliefs about one's self as a mathematics learner, one's perceptions of how others perceive him or her as a mathematics learner, beliefs about the nature of mathematics, engagement in mathematics, and perception of self as a potential participant in mathematics (Solomon, 2009).

Ratios and Proportional Reasoning

Throughout the novel, there are references to characters' sizes. For example, Stanley Yelnats is described as overweight (p. 7). Teachers can introduce Da Vinci's Vitruvian Man and then use body parts that are proportional, rather than using weight, to do comparison ratios. Here are some safe comparisons:

- The height of the "face" is about equal to the length of the hand ("Body Proportions," 2021).
- The length of a person's outspread arms is equal to their height.

- From the chin to the top of the forehead and the lowest roots of the hair is a 10th part of the whole height of a person.

Teachers can ask students to make a chart with the measurements of the different body parts mentioned above and then ask the students to compare the related body parts (see Table 3.2 for sample measurements). Rich discussion can be had around the different ratios and what they represent. The students, for example, can talk about factors that lead to the aforementioned proportions not being true for all people.

In Table 3.2, Person 1's measurements were close but not quite an inch-to-inch match. The person had long arms and a short torso, and thus there was not an inch-to-inch match between the height and the outspread arms. Person 3's outspread arms and height were the same. Where measurements are close, students might infer that there may have been measurement error. Teachers can lead a discussion about unit rates and what they mean.

Another example of people being compared in the novel occurred when Stanley first arrived at Camp Green Lake, he reflected on the power dynamics within the group of campers as he compared their sizes (p. 53). Since no specific measurements are given, teachers can organize students into pairs or small groups and ask them to record the strategies they would use to place X-Ray, Zero, Armpit, and Zigzag in order of height. Students should be expected to use the excerpt from the text to justify their placement. Based on the information in the excerpt, students can infer that the campers, in order from tallest to shortest, are Zigzag, Armpit, X-Ray, and Zero.

After engaging students in a whole-group discussion about their strategies, teachers can then provide each group with one of the Height Cards (Table 3.3).

Table 3.2 Sample Body Measurements

Person	Head Height	Hand Length	Ratio of Head Height to Hand Length	Length of Outspread Arms (Fingertip to Fingertip)	Height	Ratio of Outspread Arms to Height
Person 1	8 inches	7.5 inches	8/7.5	69.5 inches	65 inches	69.5/65
Person 2	8.5 inches	8.5 inches	8.5/8.5	75 inches	71 inches	75/71
Person 3	7 inches	6.5 inches	7/6.5	62 inches	62 inches	62/62
Person 4	7.5 inches	6.5 inches	7.5/6.5	62.5	63 inches	62.5/63

Table 3.3 Height Cards

66 inches	68 inches	70 inches
72 inches	74 inches	76 inches

Have students use one or more of the strategies shared in the earlier discussion to answer the following question: *If the tallest camper were the height shown on your height card, and the shortest camper was 5/6-foot shorter than the tallest camper, how tall could each of the boys be? Explain your reasoning.*

As students attend to the meaning of each measurement to determine reasonable heights for each of the campers in this situation, they will also need to use ratio reasoning to convert inches to feet or feet to inches. For example, students could use a representation such as a table, tape diagram, double number line, or equation to demonstrate that 5/6 of a foot is equivalent to 10 inches, so if the tallest camper is 66-inches tall, the shortest camper would be 56-inches tall.

There is also an opportunity for students to engage in mathematical modeling as they discuss rates related to the speed at which different boys dug holes, and the factors related to their speeds by engaging with this *Digging Task*:

> In Chapter 7, Stanley starts digging his first hole in the morning while the sky was still dark (p. 26). By the time the sun was rising over the horizon, Stanley dug a hole that was 3-feet deep in the center (p. 31). *How much longer might it take him to finish digging his hole?*

First, teachers and students should determine the time at which Stanley might have begun digging the hole. Invite students to research sunrise times by region and time of the year, then use this information to estimate the time Stanley started to dig his hole. As this activity supports authentic mathematical modeling, more than one answer is possible. Another assumption that will need to make before calculating the rate will be the width of the hole at sunrise. From the text, students know the hole is three-feet deep in the center, but the width is not given. Students would then take this information to determine the volume of the hole, the digging rate(s), and the time Stanley might have finished digging the hole. A sample solution is provided in table 3.4.

Encourage students to use pictures, diagrams, graphs, or tables to model these situations, compare strategies used to identify the rate, and determine

Extracting Mathematical Topics Embedded in Holes   53

Table 3.4 Sample Solution to Digging Task

Time	Dimensions of Hole (in Feet)	Explanation
4:30 AM	0×0	Stanley starts digging the hole in the morning while the sky is dark. We determined that he starts digging at 4:30 am.
6:00 AM	2×3	Volume of a cylinder = Area of the base×height $$V = \pi r^2 h$$ $$= \pi(1)(3)$$ $$= 9.42 \text{ cubic feet}$$ We found that the sun rises at 6:00 am, and also assumed that the width of Stanley's hole would be 2 feet, which means that he removed 9.42 cubic feet of dirt. Given this information, we calculated that Stanley is digging the hole at a rate of 6.28 cubic feet per hour. $$\frac{9.42}{1.5} = 6.28 \text{cubic ft/hour}$$ Volume of a cylinder = $\pi r^2 h$ $$= \pi(2.5)(5)$$ $$= 39.27 \text{ cubic feet}$$ One hole contains approximately 39.27 cubic feet of dirt, so we divided by the hourly rate to determine it would take about 6.25 hours to dig the hole. $$39.27 \div 6.28 = 6.25 \text{hours}$$ If he digs at a constant rate, Stanley should be done digging this hole by 10:45 AM. If he stops to take a break, then he might finish digging by 11 or 11:30 AM.

proportionality. As students explore the concept of rate using multiple representations, they develop a deeper understanding of proportional relationships (NCTM, 2014).

Measurement and Geometry

Given that the boys had to dig holes that are five-feet deep, and five feet across in every direction (p. 13), teachers can ask students a variety of measurement questions. Below are some possible math tasks that range in difficulty from lower-level calculation problems to higher-level multistep problems:

- What is the diameter of one of the holes? Since the diameter of a circle is any straight-line segment that passes through the center of the circle and whose end points lie on the circle, and we are told that the holes were

five-feet wide in every direction, students can determine that the diameter of each hole is five feet.
- What is the circumference of one of the holes? Circumference is the perimeter of a circle or the measure of the distance around the circle. The formula for finding the circumference is $2\pi r$ or πd where r is the radius of the circle and the diameter is equal to 2 times the radius. Since we know that the diameter of each hole is five feet, the circumference of one of the holes is 5π feet or approximately 15.7 feet.
- What is the area of the base of one of the holes? The area of the base of one of the holes is the area of a circle, πr^2. Since the diameter of one of the circles is five feet, we know the radius is 2.5 feet. So, the area for the base of one of the holes is approximately $3.14 \times (2.5)2 = 19.625$ feet^2.
- If you knew the area of the dried-up lake, how could you determine the possible number of holes that need to be dug to cover the whole lake? One could divide the area for the base of one hole into the area for the lake to get a close estimate for the number of holes that need to be dug.
- What is the volume of one of the holes dug by the boys if they followed the directions that they were given for digging the holes? We can infer from the text that the holes dug by the boys are cylinders. The formula for the volume of a cylinder = $\pi r^2 h = 3.14 \times (2.5 \text{ feet})^2 \times 5 \text{ feet} = 98,125$ feet^3.
- How did the boys know when their hole was five-feet deep? According to the novel, the shovel was five-feet long from the tip of the steel blade to the end of the wooden shaft, so the boys could use it as a nonstandard measuring tool.
- How much dirt would have been in 10 holes? Students can find the volume of 1 hole and then multiply their answer by 10. Teachers may orchestrate discourse and facilitate students' solving of this question by asking the following questions:
  - What is the question asking you to find?
  - What do you know that can help you to solve the question?
  - What strategy can you use to solve the problem? How much dirt would have been in one hole?

Number and Operations

Mr. Pendanski introduced Zero to Stanley by saying "there's nothing inside his head" (p. 19) and later diminishes him by saying he's "not completely worthless" (p. 58). However, like the character, Zero, in this book, the number zero (0) is useful and important in mathematics. Have students work in small groups to create an oral presentation or dramatization to explain the importance of the number zero. Using unconventional tools for communicating about mathematics invites students to broaden their approach to mathematical reasoning. Some mathematics topics students may consider as they are developing their presentation include (1) operations with zero (adding, subtracting, multiplying, and dividing), (2) zero in ratios and fractions, (3) zero and the additive inverse, or (4) zero in multi-digit whole numbers and decimals.

Teachers may also reference the excerpt (p. 99) about Zero (the character) from the dialogic journal to make connections between how Zero's identity as a character was formed and how the identity of the number zero is formed. There are certain characteristics of the number zero that make some operations on it undefined. For example, division by zero is undefined. In many ways, the character Zero could also be described as undefined, since his actions were not representative of who he truly was. Another important fact about the number zero is that it is the mathematical point of reference for determining the additive inverse. When describing the opposite of a number, students should understand that this value is determined by its distance from zero in the opposite direction on a number line. Additionally, zero is more than a placeholder. When zero is a factor in a multiplication equation, the product will always be zero. Students can be asked to consider these characteristics of the number zero and relate them to the character, Zero, in their dialogic journal.

Palindromes

A *palindrome* is a word or sentence that reads the same forward as it does backward. Stanley Yelnats's name, for example, is spelled the same frontward and backward. In the novel the author stated, "Everyone in his family had always liked the fact that 'Stanley Yelnats' was spelled the same frontward and backward" (p. 9). Teachers may ask students to find three-digit or four-digit numbers, such as 323, 434, and 1,991, that read the same forward and backward. This activity can be done quickly during the reading of the story. Palindromes often receive special attention in mathematics when the calendar year is a palindrome or when a particular date is a palindrome. For example, January 20, 2021 (1202021), Inauguration Day, was the first palindrome-number Inauguration Day in American history. Students may find it interesting to look for other historic palindrome dates. Students often enjoy finding palindromes due to the symmetrical nature of the arrangement of the digits. If students really get excited by palindromes, teachers can find several activities to use from *NCTM's Problems of the Week* on their website under classroom resources. The extra activities can be completed after the novel is finished.

AFTER READING *HOLES*

After reading *Holes*, students can conduct an in-depth thematic analysis as a way to use all they have learned from the text and apply it in two ways. Drawing on our theme of equity and power dynamics, students can (1) use textual evidence to trace the development of the theme across the novel and (2) make relevant connections between the novel theme and the ways it manifests in the world around us. Students can build on the dialogic journal entries they posted while reading to complete this activity. While teachers can determine the best way for students to articulate their analysis (e.g., essay, podcast, multimodal), it is important to ensure rigor and deep, meaningful analysis. As such, several prompts to guide students as they begin their analytical work can prove useful. What follows is a list of helpful questions, along with examples of textual references they might select and responses they might provide to each question.

Textual Examples	Sample Responses
The adults charged with caring for the boys use their positions to force them into manual labor—hole digging in this case—in an effort to enrich themselves (examples found throughout the novel).	I feel like we see similarities in a variety of spaces: criminal justice, employment driven by capitalistic greed, and even in education like for profit colleges. Too often, adults have the power to treat adolescents as they see fit.
Pendanski and Mr. Sir, and more indirectly the Warden, work to convince the "campers" that they are responsible for their own current situations—that is, that they deserve what they are getting (e.g., pp. 13–14, 17, 57–58).	We definitely see this in school. Students are blamed for failing grades, called lazy, labeled as "struggling" or "at risk." Instead, schools and teachers should be reflecting on the ways they fail students.
After the metallic tube engraved with the letters KB was found, the Warden had the boys stop digging and sort through the dirt. This was completely contrary to the reason they were told to dig. (pp. 64–71)	Kids get this run around all the time. Whether it's parents, teachers, or just adults in general, their intelligence is often insulted by assuming they don't recognize that they're being lied to.
Pendanski, at the order of the Warden, and in an effort to cover their own crimes and transgressions, tried to remove records of Zero from the state database (p. 144).	Students who teachers and administrators believe will fail important tests and hurt the school's rating are sometimes told to stay home on the days of testing, effectively erasing their existence as students—or they are pushed out of school in a variety of indirect ways. And more generally, schooling and society seem to want to remove who the students are as young, developing humans and replace those identities with academic knowledge, maturity, and other things adults believe youth need.

What examples do we see in the novel of adults using their own position/ authority over youth to benefit themselves? How do similar instances play out in the world around you?

Textual Examples	Sample Responses
Because Stanley was from a poor family, he did not have adequate legal counsel, never mind that he was innocent, and he believed a "camp," something he had never before seen, must be a better place (p. 5).	Considering Stanley's situation can help us think more about the ways money and wealth influence the criminal justice system. It also begs the question, how many of those in jail/prison are there because of their socioeconomic situations and not because of something they did?
The boys at the camp are a diverse group (Sachar uses the identifiers black, white, and Hispanic), and while each has tried to take some power for themselves by taking up nicknames, Pendanski refuses to see them on their own terms and instead says he intends to use the names society recognizes (p. 18).	This example from the novel forces us to interrogate how adults construct adolescence—that is, how they have the power to determine who young people are and who they can become.
The Warden, a white woman, controls the lives of diverse youth in an effort to enrich herself (seen throughout the novel)	Analyzing the Warden and her motivations and her position of authority provide opportunities to question instances of capitalism, the racial wealth gap, echoes of slavery and indentured servitude, and so forth in our society.
While told in third person, the novel centers Stanley's and his families experiences as the focal point—for example, Stanley finds the artifact that gets X-Ray a day off (pp. 64–71), teaches Zero to read (e.g., p. 99), covers for the boys by taking blame for stealing the sunflower seeds (pp. 88–91), and saving Zero's life (pp. 173–181).	Even though Stanley is ultimately mistreated and experiences injustices in his own way, he also represents a range of privileges and social problems: the white, male gaze; white privilege; the white savior trope, a focus on white as "normal" against which all else is measured, etc. In other words, a nuanced analysis of Stanley can also help us analyze race and white supremacy around us.

In addition to the murder of Sam, what evidence do we see of on-going racial, ethnic, cultural, and/or class-based injustice throughout the novel? And in what ways does that help us to more critically consider our world?

Example of Student Response Excerpt
It seems that everything about Camp Green Lake, from its historical connection to a racist murder to its connection to the State of Texas criminal justice system, is problematic and dangerous. The camp itself is run by a descendant of Trout Walker, a once wealthy (pp. 102-103) and racist man who murdered a Black man for kissing a white woman (p. 115). This, and the greed it brings about, leads the Warden to use her position and authority to mistreat the boys. We know from research that modern prisons are descendants of enslaved labor and that criminal justice policies and law are built upon notions of white supremacy. Moreover, the continued privatization of the prison industry builds wealth for a few on the backs of the oppressed and taken advantage of. In the novel, the boys are told to get used to being thirsty (p. 15). They are served terrible food (e.g., p. 21, 104). And those in charge assume they have complete authority and freedom to do whatever they want. In short, those who are sent to Camp Green lake have their humanity taken away. We see echoes of this in our society as well. Just as the Warden, Pendanski, and Mr. Sir believed they could simply delete Zero's existence from the system (p. 144), we have seen multiple examples of young people, especially young People of Color, disappearing into the justice system. Whether there is a similar lack of oversight in our juvenile and adult detention centers, very little information appears to be public knowledge, which can create opportunities for wrongdoing. And similar to Camp Green Lake, there has been plenty of reporting on poor conditions in prisons and detention centers; we've seen more and more of this in the news as a result of the pandemic, for example.

What makes Camp Green Lake problematic, and even dangerous, for its "residents"? And what policies, procedures, and systems create similar issues for other juveniles incarcerated in the United States?

Regardless of the ways students choose to share their findings, the overall goal of the thematic analysis, as from textual analysis itself, is to apply the critical lens(es) they have developed as they read the novel to all they encounter in their worlds.

BEYOND *HOLES*

As an extension to the novel, teachers can have students build on their reading of *Holes* and study of power and inequity and apply what they have learned in disciplinary ways. One way students can extend their work is through the *Turning Your Lens on the World* activity. The stated aim of this activity is for students to take what they have learned from their reading and apply their critical literacy skills and lenses to the world they inhabit, specifically to an issue of inequity they would like to address.

Students first select the issue they would like to explore. As a way to scaffold this for students, teachers can work with the whole class to generate a list of the inequities and injustices they have discussed in relation to the novel and those they made connections to in the thematic analysis activity. While not comprehensive, such a list could include racial and/or gender issues within the juvenile justice system, policies and practices at their school (e.g., dress codes, tracking, discipline referrals, and outcomes), manifestations of the racial wealth gap in their communities, disparities in educational attainment, and so on. The possibilities here are near endless, and teachers can encourage students to think outside of the class generated list to select their own issue, but providing time in class to review examples can serve as a fertile starting point for many students. What follows are some suggestions for where students can find information on these example topics.

Racial and Gender Issues within the Juvenile Justice System. Southern Poverty Law Center website has a searchable database of articles on many social justice and equity topics, including those relating to race and gender in juvenile justice. "Juvenile Justice Statistics" within the U.S. Department of Justice website. The juvenile justice statistics provided by the U.S. DOJ offers national report bulletins highlighting race and gender statistics in juvenile justice.

School Policies and Practices. Educational Leadership and ASCD. ASCD, via Educational Leadership, provides ideas and vignettes for students to consider ways to disrupt inequity within their schools. "Addressing Inequities in School Policies, Policing, and Discipline Practices," part of Professional

Learning and Community Engagement (PLACE) at the University of Wisconsin, Madison. The PLACE website hosts the "Real Talk for Change" symposia/video series featuring UW-Madison faculty and area school personnel.

Racial Wealth Gap in Communities. "A Conversation about the Racial Wealth Gap—and How to Address It" from the Brookings Institute. The Brookings Institute website hosts an informative blog entry, including links to videos and other related research around the racial wealth gap. "The Growth of the Suburbs—and the Racial Wealth Gap," a Lesson Plan for Teachers from PBS. The PBS website includes a lesson plan for teachers on the racial wealth gap and the growth of the suburbs. The lesson plan includes activities, example data for students to analyze, and links to other resources.

Disparities in Education Attainment. Ethnic and Racial Disparities in Education: Psychology's "Contributions to Understanding and Reducing Disparities" from the American Psychological Association (APA). Although this APA executive summary is intended for professionals, it is organized and articulated in a student-friendly format. The authors of the executive summary break down disparities, advocacy, educational practice, and research into digestible bullet points and numbered lists. "Equity of Opportunity" from the U.S. Department of Education. The U.S. DOE website reports the challenges and successes they perceive in impacting educational equity.

Students, then, conduct research on the issue they select, using questions like those listed below to guide their inquiry:

- When and how did the issue originate?
- In what ways does it manifest itself in society?
- Whom does it impact?
- Does it impact everyone equally and similarly?
- What statistics and other quantitative information on your issue are available?

After conducting research and using their critical lens to interrogate the data that they locate, and their own evolving understandings of the inequity itself, students can publish and share their findings with classmates. One way to make this interactive and engaging is to create a gallery walk, a sort of living inquiry installation. Rather than composing traditional research reports or presentations to be shared in front of the entire class, teachers can help students create their own exhibits consisting of posters, images, interactive information, and/or dynamic media. Students can set up their exhibits around the classroom or other school space and then take on the role of curator by welcoming their peers to their installation, introducing the information they will encounter, and answering any questions they have. Teachers may also go a step further and host the gallery walk after school and invite parents and families to attend. Regardless of the method of articulation and sharing, it is also vital students have an opportunity to reflect on their experiences and learning. Following the gallery walk, teachers can ask students to spend time engaged in reflective writing around two prompting questions:

- In what ways did applying a critical literacy lens to the issue help you to better understand it and to consider possible solutions?
- In what ways does applying mathematical ways of thinking to the issue, including the information you found as part of your research, help you to better understand how and why the issue exists (and how and why it impacts different people differently)?

CONCLUSION

In this chapter, we shared how the YA novel *Holes* can be used as a catalyst for helping students to think about social issues, while simultaneously using mathematics as a critical tool to examine the issues. Moreover, we offered activities that could be used as a means to help students develop essential mathematics skills like probability, proportional reasoning, statistical reasoning, and measurement. We also shared how multiple entry levels and high cognitive demand mathematical tasks can be extracted from the plot of the novel, all of which can aid in students developing mathematical habits of mind, such as making sense of problems and persevering in solving them, reasoning abstractly and quantitatively, constructing viable arguments and critiquing the reasoning of others, attending to precision, and others (National Governors Association and Council of Chief State School Officers, 2010; NCTM, 2014). In addition, we used the notion that "Stanley Yelnats" is a palindrome and asked students to think about numbers that are palindromes—doing so could lead students to think about the puzzling aspect of mathematics and help them learn to enjoy mathematics for its own sake. We acknowledge that the activities described in this chapter are not comprehensive in nature. Instead, they are meant to offer a few ideas and extend the argument that mathematics and English Language Arts can form a powerful team for helping students to develop and apply the skills necessary for interrogating and working to interrupt inequities in our society.

REFERENCES

Aguirre, J. M., Mayfield-Ingram, K., & Martin, D. B. (2013). *The impact of identity in K–8 mathematics: Rethinking equity-based practices*. National Council of Teachers of Mathematics. Author.

Behrman, E. H. (2006). Teaching about language, power, and text: A review of classroom practices that support critical literacy. *Journal of Adolescent & Adult Literacy, 49*(6), 481–486.

Bigelow, B., Harvey, B., Karp, S., & Miller, L. (eds.). (2001). *Rethinking our classrooms: Teaching for equity and justice* (Vol. 2). Rethinking Schools.

Body Proportions. (2021, May 2). In Wikipedia. https://en.wikipedia.org/wiki/Body_proportions.

Boyd, A. S., & Darragh, J. J. (2019). *Reading for action: Engaging youth in social justice through young adult literature.* Rowman & Littlefield.

Gutstein, E. (2006). *Reading and writing the world with mathematics: Toward a pedagogy for social justice.* Routledge, Taylor & Francis Group.

Harste, J. (2003). What do we mean by literacy now? *Voices from the Middle, 10*(3), 8–12.

Janks, H. (2010). *Literacy and power.* Routledge.

Luke, A. (2012). Critical literacy: Foundational notes. *Theory into Practice, 51*(4), 4–11.

National Council of Teachers of Mathematics. (2015). *Principles to actions toolkit.* NCTM.org Online Tool. https://www.nctm.org/PtAToolkit/.

National Council of Teachers of Mathematics. (2014). *Principles to actions: Ensuring mathematical success for all.* Author.

National Council of Teachers of Mathematics. (2018). *Catalyzing change in high school mathematics: Initiating critical conversations.* Author.

National Governors Association Center for Best Practices, & Council of Chief State School Officers. (2010). *Common core state standards for mathematics.* Author.

Sachar, L. (2000). *Holes.* Yearling.

Sams, B. L., & Cook, M. P. (2019). (Un)sanctioned: Young adult literature as meaningful sponsor for writing teacher education. *English Teaching: Practice & Critique, 18*(1), 70–84.

Solomon, Y. (2009). *Mathematical literacy: Developing identities of inclusion.* Routledge.

Strutchens, M. E. (2018). Obtaining social justice via culturally inclusive mathematics. *Special Edition of the New England Mathematics Journal: Toward Teaching Mathematics Through Social Justice II, 51*(2), 18–31.

Chapter 4

Reading like a Mathematician

Exploring Data, Narratives, and Phenomena in A Long Walk to Water

Suki Jones Mozenter and Robin Keturah Anderson

Mathematics abounds in the world around us, yet students often fail to recognize its presence in their world outside the math classroom. Young adult (YA) novels can be used to help students notice the mathematics that frames their lives. Reading the YA novel *A Long Walk to Water* (Park, 2010), teachers can guide students in an exploration of perspective and scale, how parts relate to a whole, and how parts within a whole interact. Students can explore how data, narratives, and phenomena can be understood differently depending on how they are presented or how they are seen. Navigating these shifts in scale and in perspective is integral to constructing meaning out of data.

A Long Walk to Water presents multiple engagement points for readers: experiences of refugees, water rights, interpreting and presenting maps and data, and analyzing the interactions of multiple text elements. Teachers and students can connect events and issues in the text with what is happening globally and locally. *How is water security in Sudan similar to water security in Flint, Michigan? What does the refugee crisis look like globally? What does it look like in our school community? How does our understanding of the experiences of Salva, the main character and a Sudanese refugee, change when we map the distances he walked to maps of our community?* These questions are all related to how scale and perspective shape our understanding of data, narratives, and phenomena. They are also some of the questions that we suggest teachers and students consider as they engage with the text.

A key question that is central to the approaches we outline in this chapter is "What do we gain when we read as mathematicians?" In each section, we offer prompts and strategies for engaging students in metacognitive reflection on their identities and sensemaking as they read. We present this mathematical and literacy integration as a pathway for students and teachers to explore and develop the stance of reading as a mathematician. *How does our interpretation of, and connection to, the narrative change when we read as mathematicians? How do our identities as mathematicians influence our identities as readers and vice versa? How do we already read as mathematicians and what other opportunities do we have to develop this reading lens?*

A LONG WALK TO WATER BY LINDA SUE PARK

A Long Walk to Water, by Linda Sue Park, is a fictionalized account—based on factual events—of the impacts of civil war and clean water access on Sudanese youth. The text follows Salva, a character based on the real-life experiences of Salva Dut, a "Lost Boy," as he flees the impacts of the 1985 Sudanese Civil War. He is first separated from his family, walks across multiple countries to flee the war, and eventually finds himself stuck in a series of refugee camps. He resettles in the United States and later returns to Sudan to reconnect with his family and begin a safe water access project. During the narration of Salva's experiences, the author intersperses the experiences of a young Sudanese girl, Nya, and her family as they struggle to secure clean water in 2008 Sudan. Nya's story intersects with Salva's when Salva's organization builds a drinking well in Nya's village.

BEFORE READING *A LONG WALK TO WATER*

The time before reading a text gives teachers and students an opportunity to begin to explore ideas of reading as mathematicians, as well as the concept of scale, while also orienting themselves to *A Long Walk to Water*. In this section, we describe the launch of the book, the consideration of scale, and the idea of reading as a mathematician.

BOOK WALK: ELEMENTS OF THE TEXT
AND WHAT THEY TELL US

On the day the novel is introduced, have students engage in a book walk. A book walk is a process of self-guided discovery by which readers begin to orient themselves to the text and the reading. Discuss with students what they do when they are choosing a book to read. What do they look at? Some students may offer that they look at the image on the front cover, or read the description on the back cover, or that they look inside the back cover for information about the author. The teacher can build on the text elements suggested by students and offer others, including table of contents, front material, images, illustrations, or unusual text elements within the book.

A Long Walk to Water presents a number of interesting text elements to consider during a book walk: a note on the front cover that it is "based on a true story"; depending on the edition, a note on the front cover that it was a *New York Times* bestseller and that the author was a Newberry Award Medalist; two short descriptions on the back cover that suggest two different stories; an author photograph that features two people, the author and the person whose life the story is based on; a map in the front material; two different font styles within the book; dates and locations listed at the beginnings of the chapters. Teachers can engage students in small- or large-group discussions around what elements they notice and how they are making sense of the elements themselves and the book.

This is also a good time to introduce an interactive notebook. The notebook is a place of reflection and sensemaking for students, and it can include writing, drawing, or calculating. It can also be a place for teachers to provide feedback and respond to student work. As students read, they can use their notebook to respond to the text, work out problems of scale, and reflect on the impacts of reading as a mathematician.

READING AS A MATHEMATICIAN: AUTOBIOGRAPHICAL KEY MOMENT

During whole-class discussion, the teacher could highlight that as we read, we bring many different identities to what we are reading. We read as students, or people who love animals. Or we read as LGBTQ+, or white, or black, or Latinx readers. We bring many different identities to who we are as readers. As students work through this text, they can highlight what it means for them to read as mathematicians. Have them pause and bring their identities as mathematicians to read as mathematicians.

To prepare students to read as mathematicians, start by asking them to reflect on their own identities as mathematicians. Invite students to think about key moments in their development as mathematicians. *Were there particular experiences that shaped how they see themselves as mathematicians? Were there people who had a significant impact?* Model this by recounting to students a key moment in your own development as a mathematician. As you recount the moment, use a combination of sketching and key words to relate it. Then, invite students to think about a key moment in their mathematical autobiographies. In their interactive notebooks, have them sketch and/or write a description of this moment. You could then give students an opportunity to recount their key moments in small groups.

After students have reflected on, recorded, and shared their key moment, have them theorize in small groups what it might mean to read as a mathematician, how they might do that, and how that might impact their experiences as readers and how they make sense of what they read. Following the small-group discussion, invite students to share what they discussed. Use their responses to start a class list of what it means to read as a mathematician.

INTRODUCING MATHEMATICAL SCALE AS AN INTERPRETIVE TOOL

Prior to reading, the teacher can leverage the map of Salva's journey (Park, 2010, p. x) to introduce the mathematical concept of scale. The concept of scale "refers to the spatial, temporal, quantitative, or analytical dimensions used by scientists to measure and study objects and processes" (Gibson, Ostrom, & Ahn, 2000, p. 218). Through the discussion of maps as spatial representations of scale, students can interpret the length of Salva's journey. The following questions can be used to guide the class discussion:

- What do you notice about the map?
- How can we tell this is a scale drawing of Central Africa?
- What part of the map indicates Salva's journey?
- How can we use the map to estimate the length of Salva's journey?
- What tool(s) could help us get a more accurate estimate?
- How can maps be useful?

Once the students have built a collective understanding of how scales are represented and used to investigate visual models, they can turn to their local community to interpret the length of Salva's journey. This aspect of the lesson continues to build a whole-class working definition of scale. Teachers can create a printable map of their context. Figure 4.1 provides an example of a usable map created in Google by the authors. The star was added to represent the school's location and starting point for student exploration. The provided scale (five miles) was recreated to be visible, as the Google map's provided scale is small and located in the bottom right corner of the map. It is also recommended to create a printable map with a different scale than the novel map to discuss and compare different scales.

As students engage with exploring a new scale representation, the students could investigate the following questions. It should be noted that students might struggle mapping Salva's journey on the provided contextual map;

Reading like a Mathematician 67

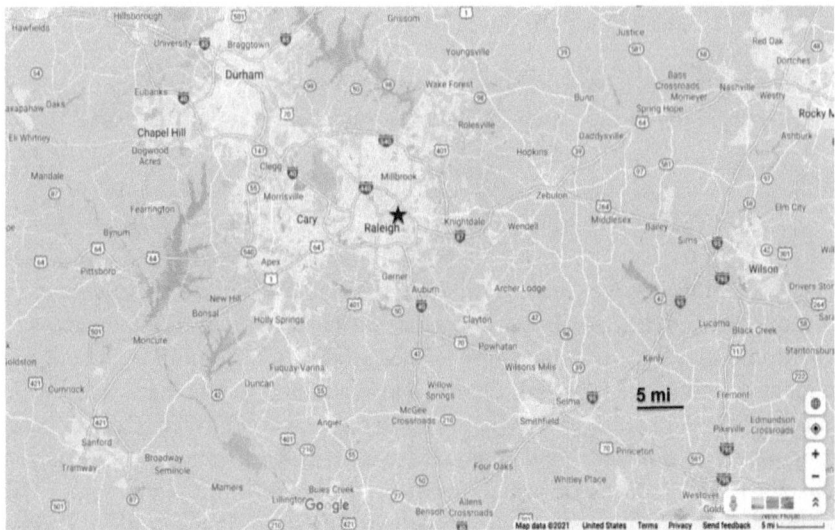

Figure 4.1 **Example of a Printable Map.** *Source*: Created by authors in Google

this should surface discussions based around the magnitude of his journey in relation to the students' communities. Teachers might encourage students to generate their own maps to explore Salva's journey in contexts that they are familiar with:

- How far away from school could you travel if you walked the same distance as Salva?
- How does the scale of the novel's map relate to the scale on our map?
- What similarities and/or differences do you notice between the two maps?
- (Extension.) How does requiring the usage of maintained roads change your final location? (Students can explore the usage of maintained roads on map sites like On The Go Map).

After using the mathematical concept of scale to simulate the length of Salva's journey within their local community, students can reflect on how reading as a mathematician might enhance their understanding of a text. Teachers can use the following prompts as exit tickets, discussion prompts, or written reflections in their interactive notebooks:

- How does the application of scale to a relatable context impact your predictions about what we will be reading?
- In what ways did your predictions change when you mapped Salva's journey onto our local map?

- What new questions or wonderings arose for you as we worked with our local map?
- What other texts have you read where a map was part of the story? In those instances, how did the map impact your understanding of the story?

WHILE READING *A LONG WALK TO WATER*

The book unfolds two parallel narratives: Salva's life as a refugee and, decades later, Nya's precarious life as a Sudanese girl with limited access to safe drinking water. In this section, we describe two focal moments to engage students in concepts of scale, extent, and reading as mathematicians, while also exploring issues of water scarcity and refugee conditions. Each moment contains three different mathematical inquiries. The first moment is during Chapter 4 and focuses on water scarcity, Nya's story, rate, and scale. The second is during Chapter 11 when Salva arrives at the Itang Refugee Camp; this moment uses global refugee data to explore the following question: *How do we know whether a number is large or small?*

Chapter 4: Nya's Daily Water Journey

At the beginning of the first four chapters, Park describes Nya's daily ritual of securing clean water for her family (pp. 1, 8, 14–15, 20). Nya spends most of her day making two trips to the water source with just a brief rest for lunch. To help understand the impact of securing clean water on Nya, students can use mathematics to quantify Nya's journey by calculating a theoretical walking speed for Nya, calculating the distance she could have traveled at that rate, then interpreting Nya's experience through the context of their local community by using scale to map Nya's journey to the student's local community. This activity leverages the map analysis skills acquired in the before-reading activity.

Part I: How Fast Did Nya Walk?

Begin by asking students how fast they think Nya would have walked when she went to get water for her family. This question should provide students time to guess her speed using their own experiences and also offer an opportunity for students to identify variables that they would need to know in order to be able to make more precise predictions. These variables might include the distance she is walking, the weight she is carrying, and obstacles she might encounter.

Once guesses have been charted, students can gather data to make a class prediction. To collect data, students need a stopwatch and measured distance (e.g, a track, a long hallway that has been measured) where they can time themselves as they walk. Prior to releasing the students in small groups to collect data, the teacher should lead a class discussion about relative pace for the data collection. The following questions can be used to guide the conversation:

- What are some factors that might impact Nya's pace (pp. 8, 14–15, 20)?
- How might we account for these factors when estimating Nya's pace?
- What would happen if we did not take these into account? Why might it matter?

In small groups, students can calculate their walking speed. They should note how many seconds it takes them to walk the predetermined distance. After all students collect their data, the group should average their times to get one speed (feet/second). The average YA walks 4.09 feet/sec (Knoblauch et al., 1996). This speed could be used by groups or classes in which collecting data is not an option. The students can then work together to convert the speed from feet/second to miles/hour.

Because half of Nya's journey is completed while carrying close to 40 pounds of water, the students should also calculate a slower walking pace for themselves. This should be done by using a scale factor of 0.75. It can reasonably be assumed that a person's walking speed will decrease by 25% due to the added weight. The students should now have two walking speeds: one for walking without weight and one for a theoretical speed for walking with a weight.

While it may be appealing to have students replicate Nya's return with the water by having them carry jugs of water themselves, we would caution against this. There is a delicate tension when engaging in simulations. The tension is between building students' understanding or empathy and minimizing or othering the lived experiences being simulated. For this reason, we encourage you to have students collect data on their unburdened walking pace and then use scale factor to calculate a burdened walking pace. Beyond the mathematical extension that this approach provides, it is also intended to limit the negative repercussions that such simulations could spark.

Part II: How Far Did Nya Walk?

With the calculated speeds and information from the book about Nya's two daily trips to collect water (e.g., left at breakfast and returned at lunch [pp. 1, 14–15, 20]), traveled to the pond twice daily (pp. 1, 20), the students can calculate the total distance Nya traveled in one day to secure clean water for her family. As students are working in small groups to predict the length of

Nya's journey, the teacher could monitor with the following probing questions to advance student thinking:

- How does your time walking compare to Nya's time walking?
- How much time do you think Nya spent walking to the water source versus walking home from the water source?
- How can you account for the trip before lunch and after lunch?

The intention is for these questions to be used by the teacher as probing or "back-pocket" questions while monitoring small groups. If desired, however, these monitoring questions can become reflection questions during a whole-class debrief.

Part III: Mapping Nya's Walk within the Local Community

After students have calculated a distance for Nya's daily journey, they will bring this distance into their local context by using the scale on a map of their local community. Mapping Nya's journey within their local community—like the previous mapping activity—provides students a contextually grounded experience for mathematical meaning-making. The printable map should include the geographic region around your local community. Keeping the map local will allow students to use known landmarks to help interpret the scale of Nya's journey.

As students conclude this three-part examination of Nya's journey through the mathematical tools of rate, scale factors, and scales, engage students in a reflection on how reading like a mathematician has impacted their understanding of Nya's experiences. This reflection can be done in small groups and/or independently in their interactive notebooks. Some prompts that you might use are as follows:

- In what ways did calculating and mapping Nya's journey help develop your understanding of the story?
- In what ways did calculating a walking rate develop your understanding of Nya's walk for water?
- What insights do you have about Nya's journey after mapping it?
- In what ways has reading as a mathematician developed your understanding of Nya's life?

Following individual or small-group consideration of these prompts, revisit the earlier list of impacts of reading as a mathematician. Invite students to revise any previous comments and to add to the evolving class list.

Chapter 11: The Refugee Crisis

This while-reading activity provides students ways to use the quantitative and analytical dimensions of scale to meaningfully interpret the number of Sudanese people that entered Ethiopian refugee camps. As they analyze statistics relating to refugee camps and the refugee crisis, students will be investigating a central question: *How do we know when a number is big or small?*

Salva entered the Itang Refugee Camp in 1985 shortly after the death of his uncle. The author described the number of refugees in the camp by asking, "How could there be this many people in the world?" (p. 66). Continuing to build an image of overcrowding, the author described the physical environment, "People in lines and masses and clumps. People milling around, standing, sitting, or crouching on the ground, lying down with their legs curled up because there was not enough room to stretch out" (p. 66). To support the author's description of the camp, teachers are encouraged to show a three-minute documentary produced by the United Nations High Commissioner for Refugees (UNHCR) that depicts the influx of refugees to a similar Ethiopian refugee camp (UNHCR, 2014).

To better understand the number of Sudanese refugees in Ethiopia, students can use the scale concept of *extent*. Extent refers to the "magnitude of a dimension" and "fixes the outer boundary of the measured phenomenon" (Gibson, Ostrom, & Ahn, 2000, p. 219). When referring to locations, such as Ethiopia or Sudan, extent can help bound the scale to geographical regions with small scale (one refugee camp), a medium scale (the country of Ethiopia), or a large scale (the world). The teacher, and students, should feel free to change the extent of their interpretations to help justify their claims throughout this activity.

The central question for this activity is "How do we know when a number is big or small?" The first part of the activity explores what information students need to decide whether a number is large or small. In Part II, students will be asked to justify whether a given number of Sudanese refugees in Ethiopia is large or small by exploring density within refugee camps (small scale). Part III will have students explore whether the number of Sudanese refugees in Ethiopia is big or small by investigating proportions of refugees (medium and/or large scale). Finally, the students will bring their analysis of the extent of scale to the United States by beginning to think about refugees within their communities.

Part I: What Information Is Needed to Determine Size?

Students should use different levels of scale to build an awareness of the number of people entering Ethiopian refugee camps due to the civil war in Sudan. To begin this activity display figure 4.2 from the UNHCR Refugee

Year	Country of Origin	Country of Asylum	Refugees under UNHCR's mandate
1985	Sudan (SDN)	Ethiopia (ETH)	180,000
1986	Sudan (SDN)	Ethiopia (ETH)	132,000
1987	Sudan (SDN)	Ethiopia (ETH)	250,000
1988	Sudan (SDN)	Ethiopia (ETH)	328,976
1989	Sudan (SDN)	Ethiopia (ETH)	384,989
1990	Sudan (SDN)	Ethiopia (ETH)	387,585
1991	Sudan (SDN)	Ethiopia (ETH)	15,670

Figure 4.2 Sudanese Refugee Population in Ethiopia 1985–1991. *Source*: UNHCR Refugee Data Finder, 2021

Data Finder (UNHCR, 2021). Ask students to reflect individually in their interactive notebooks using the following prompts: *What do you notice about the statistics in the table? What do you wonder about after reading this table?*

Once students have some time to individually think, the teacher could lead a class discussion around the students' noticing and wonderings. The teacher should be aware that the sharp drop in refugees in 1991 corresponds to the same time that Salva was kicked out of the Itang Refugee Camp. This was a time of government transition at the end of the Ethiopian Civil War. The novel describes the forced removal of the refugees in Chapters 12 and 13.

After initial reactions to the statistics, shift the focus to the number of Sudanese refugees in Ethiopia in 1990. The teacher could pose the following question: *Is this a small or large number of people?* This question allows for a deeper analysis of one number. While 387,585 does represent the number of Sudanese refugees in Ethiopia, it also opens up opportunities for larger discussions of the scale of the refugee crisis not just in Sudan and Ethiopia, but also across the continent of Africa and the world. After the students have initial reactions, the teacher could ask, "What information do you need to decide if this number is large or small?" The teacher should record a list of desired information to use in the next part.

Part II: Refugee Living Situations: A Small-Scale Interpretation of a Number

The refugee crisis in Sudan, and present-day South Sudan, has sent an influx of asylum seekers across the borders to Ethiopia for over 30 years. In fact, reports from the Gambella region of Ethiopia claim that, between the region's seven refugee camps, there were 319,130 South Sudanese refugees as of June 2020 (UNHCR, n.d.). To start a class discussion, the teacher can draw upon the list created at the end of Part I, asking students the following questions to help guide the conversation if needed:

- What information on this list can help us determine whether 319,130 is a large or small number of refugees in relation to the seven camps in the Gambella region of Ethiopia?
- What additional information would you need to justify your decision?

Students might ask for the following information: the size of an average refugee camp (2.5-square miles), the number of housing units across the seven refugee camps (40,350), the number of family units in the 319,130 people (66,898). More information is provided in the UNHCR report on South Sudanese refugees in Ethiopia if needed. Students could change the extent of the scale to different dimensions allowing them to justify whether the number 319,130 is a large or small number. If it does not arise, teachers could introduce the concept of population density—the number of individuals to a unit of area—to give the students a defined extent (the camp, the housing unit). After students have brainstormed additional desired information, provide them a class period and access to online search tools to collect it. Finally, in groups or as a whole class, students could discuss their rationale for determining that 319,130 is a large or small number.

To engage students with the issue of perspective and its impact on interpreting proportion and extent, have students reflect in a double-entry journal. A double-entry journal lists information in one column and uses the second column for student reflection or response. This could be done in students' interactive notebooks. Invite students to record in the left column evidence that they gathered to justify whether 319,130 was a large or small number. In the right column, they should then write what they think Salva's response would be to these justifications. *Would he agree or disagree? What evidence might he offer to justify his reasoning?* This could be done individually or in small groups. Close this activity by engaging students in a class discussion of how perspective might impact interpretation of proportion and extent.

*Part III: Taking Sides: Using Global Refugee
Data to Justify the Size of a Number*

To continue to guide students in the quantitative and analytical interpretations of scale, the class could move into a forced-choice activity to discuss the relative size of a number. Referring back to the number of refugees in Ethiopia at the time of Salva's tenure within the Itang Refugee Camp, students can support whether a number is large or small. In groups of 3, students can be randomly assigned to justify whether 387,585 is either a large number or a

Population of Ethiopia in 1990	47.89 Million
Population of Sudan in 1990	20.15 Million
Number of Refugees in the World in 1990	17.39 Million
Number of Sudanese Refugees in 1990	523,996
Population of the United States in 1990	250.1 Million
Number of United States Refugees in 1990	0
Number of Sudanese Refugees in the United States in 1990	44

Figure 4.3 Population and Refugee Data for 1990. *Source*: UNHCR Refugee Data Finder

small one. The teacher should draw upon the list created in Part I and also the data in figure 4.3, if necessary, to provide students information to help with their interpretation.

At the conclusion of group time, the teacher could facilitate an in-class debate, attending to how the students use proportional reasoning and extent of scale in their interpretations. The intended learning outcome of Part III is for students to realize that numbers are not inherently large or small but rather interpreted through a context.

To conclude this portion of the unit, bring students back again to consider how reading as a mathematician impacts their understanding of the text. This time, invite students to consider ways that reading as a mathematician may blind them to the unique experiences of an individual. This could be a reflection that they discuss in small groups or individually record in their interactive journals. Some prompts could include the following:

- Whose perspective should be taken into account when considering whether 319,130 is a large or small number?
- What role should mathematics play in making this determination?
- In what ways did mathematical thinking shape your understanding of the size of the number?
- In what ways did it shape your understanding of Salva's experience in the refugee camps?

Following time for individual and/or small-group consideration, revisit the evolving class list of the impacts of reading as a mathematician. Invite students again to revise the existing list and expand it. Students may also create a list of cautions to keep in mind when reading as a mathematician.

AFTER READING *A LONG WALK TO WATER*

Using Scale for a Biographical Sketch

To reflect on both the narratives in *A Long Walk to Water* and the concept of scale, students will use scale to create a biographical summary of either Nyla or Salva. Begin by having students discuss how scale can be applied to the story of someone's life. Have students brainstorm the different ways that scale could be used in a biography (e.g., the impact of events, the duration of phases). Either in small groups or individually, have students choose to create a biographical sketch of either Nyla or Salva. These sketches could be artistic, mathematical, or written. They should include an application of scale as well as an explanation of how they used scale in their biographical sketch. Students' biographical sketches can then be displayed as a gallery, and other students, families, and teachers can be invited to the gallery opening.

Using Scale to Make a Point

The while-reading activities provided students with opportunities to use scale to make sense of two social issues: safe water access and the refugee crisis. Students have used mathematical reasoning to understand the impacts of these issues and to draw parallels between the Sudanese context and their own community. This activity brings together mapping and proportional reasoning skills to prepare a product (letter, presentation, poster, etc.) that leverages the mathematical concept of scale to raise awareness of issues of safe water access.

Focusing on how scale impacts analysis, students could investigate one (or a combination) of the following ideas: (1) safe water access data globally (United Nations International Children's Emergency Fund [UNICEF], 2019; Ritchie & Roser, 2019); (2) comparing current safe water access between South Sudan and the United States (State of Michigan, 2021; UNICEF, n.d.); or (3) safe water access in the United States, identifying patterns of access and overlaying those with racial, socioeconomic, urban/rural data (UNICEF, 2019; U.S. Census Bureau, n.d.). The following questions can be used by teachers to support students' application of scale to their product:

- How do you know that a statistic is big or small?
- How much bigger (or smaller) is one statistic compared to another?
- How could using a map make your mathematical justification clearer?
- Does your data support the point you are trying to make?

The format of the product should be left open to allow for students to account for different types of data and the extent of scale that they would like to bring to the project.

BEYOND *A LONG WALK TO WATER*

Influencing Change

Based on Eric Gutstein's (2006) call for students to engage with mathematics to address injustices around them, students can take action by leveraging the product produced in the after-reading activity. Students could share their products with governmental officials, local nonprofits, or over social media to increase awareness around issues of safe water and potentially influence action in their local communities. Alternatively, students could develop a product relating to another issue relevant to this novel (e.g., refugee camps, refugee flight experiences, gender-related safe water impacts) to address a target audience.

When Is One a Big Number?

Throughout the activities in this chapter, students have been asked to interrogate the magnitude of journeys, populations, and resources by comparing quantities and relationships to scale. By making sense of these quantities mathematically, students were better positioned to relate personally to the experiences of Nya and Salva. Students have also been asked to mathematically justify statistics through understanding the extent of scale of a quantity. But understanding the quantity mathematically is not enough to have a full appreciation of a statistic. In fact, being "one of" can make one feel infinitely big. To build a sense of being "one of" ask the students to personally reflect in their interactive notebook on the following statements:

- When you're the one experiencing the situation, your experience is big even though it is just one experience.
- When one person takes action, the impact can be big.

CONCLUSION

Throughout their work with the novel *A Long Walk to Water*, students have engaged some of the tools, practices, and skills of a mathematician to make sense of critical, global issues, such as refugee experiences and safe water access. These engagements not only deepen students' understandings of the novel, Salva's story, and the issues at the center, but also provide students with opportunities to reflect on how reading like a mathematician is a skill that adds complexity and richness to their literacies practices. This experience also contributes to their developing identities as mathematicians who can use these practices to make sense of the wider world.

REFERENCES

Gibson, C. C., Ostrom, E., & Ahn, T. K. (2000). The concept of scale and the human dimensions of global change: A survey. *Ecological Economics, 32*(2), 217–239.

Gutstein, E. (2006). *Reading and writing the world with mathematics: Toward a pedagogy for social justice*. Routledge.

Knoblauch, R. L., Pietrucha, M. T., & Nitzburg, M. (1996). Field studies of pedestrian walking speed and start-up time. *Transportation Research Record, 1538*(1), 27–38.

Park, L. S. (2011). *A long walk to water*. Houghton Mifflin.

Ritchie, H., & Roser, M. (2019, November). *Clean water*. Our World in Data. https://ourworldindata.org/water-access.

State of Michigan. (2021). *Flint water*. https://www.michigan.gov/flintwater/0,6092,7-345-76292_76294_76297---,00.html.

United Nations High Commissioner for Refugees. (2014, June 3). *Ethiopia: South Sudanese refugee influx* [Video]. YouTube. https://www.youtube.com/watch?v=MMIPX0nELN0.

United Nations High Commissioner for Refugees. (2021). *Refugee data finder*. https://www.unhcr.org/refugee-statistics/download/?url=E1ZxP4.

United Nations High Commissioner for Refugees. (2020, July). *Refugees and asylum seekers from South Sudan in Gambella region: Situation Update*. https://reliefweb.int/sites/reliefweb.int/files/resources/78202.pdf.

United Nations International Children's Emergency Fund. (2019, June). *Drinking water*. https://data.unicef.org/topic/water-and-sanitation/drinking-water/.

United Nations International Children's Emergency Fund. (2021, June). *Water, sanitation and hygiene (WASH): Providing families with clear water, improved sanitation and good hygiene practices*. https://www.unicef.org/southsudan/what-we-do/wash.

United States Census Bureau. (2020). *Explore census data*. https://data.census.gov/cedsci/.

Chapter 5

Exploring *All of the Above*
Platonic Solids, Scale, Proportions, and Characterization
Jennifer R. Meadows and Amber Spears

Based on a true story, *All of the Above* (Pearsall, 2006) is told through the perspectives of a middle school teacher, Mr. Collins, and four of his students: James, Marcel, Rhondell, and Sharice. Mr. Collins and the students work together to create the world's largest tetrahedron—a pyramid whose faces are four congruent equilateral triangles. Along the way, they interact with several other characters to highlight the struggles of living in the inner city. Diversity reveals commonalities faced by those from all walks of life. The reader is immersed into the narrative through the sights, sounds, and tastes the characters experience.

Mathematically, the text lends itself to representing and finding the surface area of three-dimensional figures using two-dimensional nets, as well as using proportional reasoning to solve problems, such as converting recipes to accommodate more or fewer people. In this chapter, we have curated a sampling of activities that teachers may use to foster students' mathematical literacy through the reading of *All of the Above* as they engage in the mathematics highlighted in its pages.

ALL OF THE ABOVE BY SHELLY PEARSALL

Have your students ever worked so hard on a project just to see it ruined? Have any of your withdrawn students ever become class leaders? In *All of the Above*, you will follow the journey of an inner-city mathematics classroom turned after-school club, how the characters faced adversity, overcame challenges, and became people who believed in the power of themselves and their team. As the students worked together on their Guinness record–breaking project—to build the world's largest tetrahedron by gluing together hundreds of small tetrahedrons—they discovered that with determination and teamwork, dreams can come true.

A NOTE TO TEACHERS

In *All of the Above*, the students in Mr. Collins's middle school mathematics class express unfavorable feelings toward class participation. On page 8, Mr. Collins asked the students what would make them care about being in his math class, room 307, Washington Middle School, Cleveland, Ohio. James replied that he hated math and the entire class agreed. The students seem to believe that math is something that they are not good at and that cannot change. Carol Dweck's (2008) research would classify these students as having fixed mindsets. Before reading *All of the Above*, it may be important for teachers to help students dispel adverse feelings they may have toward mathematics and reading. Rather than settling on a fixed mindset, teachers can emphasize the power in moving toward a growth mindset, with the understanding that working through challenges helps students to press on to those goals. Following are some affirmations that students may need to hear prior to beginning this novel study: *It is okay if I cannot read all of the words in my book; it is okay if I do not understand what all the words in the book means; it is okay to ask my classmates and teachers for help with things I may not yet understand; I can tackle any mathematics problem; I can use literacy strategies to help me comprehend what I read.* Consider posting these around the classroom as encouraging reminders for readers.

BEFORE READING *ALL OF THE ABOVE*

Before reading the novel, bolster students' interest in the text by activating prior knowledge and inviting them into a new learning experience using book talks. During this phase of the reading process, teachers set a purpose for reading, introduce difficult vocabulary, and explore concepts that will be read in the text. When teachers provide opportunities for students to explore texts before they read, they allow students to deepen comprehension of the text before they begin. The following section outlines and describes pre-reading strategies that teachers can use to supplement their instruction of *All of the Above*.

Previewing the Text and Vocabulary

Introduce the text by closely analyzing the book cover: the title, the images of students' faces, the shapes, and the colors. Read the back matter. Predict why there are triangles on each of the page numbers and why each chapter is titled with a different character's name. How is this alike or different from other books the students have read? Show students that there are letters (p. 209), recipes (pp. 30, 48), and sketches (pp. 34, 58) included in the novel in addition to the fictional text. Have students make predictions about how all these features give us clues as to what the novel will be about.

In addition to previewing the text, it is essential to preview the key vocabulary students will encounter. To support comprehension before reading *All of the Above*, introduce mathematics-related vocabulary from the book. A concept map is a vocabulary graphic organizer separated into four quadrants where students write definitions based on their understanding of words, create sketches, craft sentences using the words, and connect the word to real life (Tompkins, 2004). The plot in *All of the Above* revolves around a group of students who seek to build the world's largest tetrahedron. To develop students' background knowledge about tetrahedrons before reading the novel, they could complete a concept map of the term "tetrahedron." An example of a completed concept map is provided in figure 5.1. Other key vocabulary from the novel—nets, pyramid, scale—could also be explored in this manner.

To further develop students' familiarity with the mathematics they will encounter when reading the novel, students can complete an in-depth

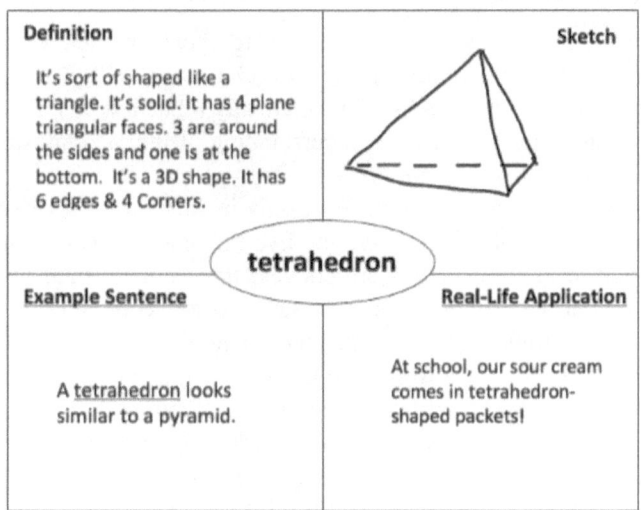

Figure 5.1 Concept Map for a Tetrahedron. *Source*: Created by authors

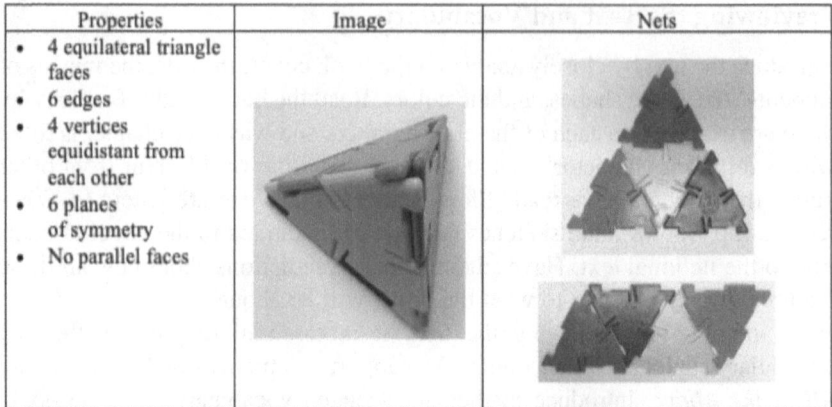

Figure 5.2 Properties of the Tetrahedron. *Source*: Created by authors

mathematical exploration of *tetrahedrons* using hands-on materials such as paper or manipulatives to construct a tetrahedron from its net (see figure 5.2). Building a tetrahedron from its two-dimensional net provides opportunities for students to analyze its mathematical structure and properties.

After students have developed familiarity with the tetrahedron, specifically, they can then begin to explore Platonic Solids more generally. This will be beneficial for students as they explore the geometric features of the three-dimensional shapes described in *All of the Above*.

Exploring the Mathematics: Semantic Feature Analysis

Another pre-reading strategy we recommend teachers complete with their students is a semantic feature analysis (SFA) (Gunning, 2020). Teachers using the SFA grid will construct a list of thematically related words. Students will use the SFA to demonstrate their understanding of the meanings of each of the words in relation to one another.

Provide each student with a grid that only includes the list of vocabulary words—in our example, we explore the five Platonic Solids of which one is the tetrahedron. Platonic Solids are convex, three-dimensional figures with identical, regular faces that are both equiangular and equilateral. Their faces only intersect at their edges, and the same number of faces meets at each vertex. Encourage students to put a plus (+) or minus (−) in each square of the grid (see figure 5.3) to indicate whether each Platonic Solid (tetrahedron, hexahedron, octahedron, dodecahedron, and icosahedron) features the characteristics listed in the top row (edges are the same length, 3D shape, triangular faces, parallel faces). The headings on the chart emphasize important information that students need in order to fully grasp Platonic Solids. The

	Edges are the Same Length	3D Shape	Triangular Faces	Parallel Faces
Tetrahedron	+	+	+	−
Hexahedron	+	+	−	+
Octahedron	+	+	+	−
Dodecahedron	+	+	−	+
Icosahedron	+	+	+	−

Figure 5.3 Semantic Feature Analysis for Platonic Solids. *Source*: Created by authors

first two headings, *Edges Are the Same Length* and *3D Shape*, draw attention to two characteristics that polyhedral must possess to be classified as Platonic Solids. The last two headings, *Triangular Faces* and *Parallel Faces*, highlight specific qualities that make specific Platonic Solids vary from each other. After students complete the grid, engage students in a whole-class debate about how the different Platonic Solids are alike and different. By completing the SFA, students should begin to see relationships among the words and they will be more equipped to apply these understandings to their reading of *All of the Above*.

Further Exploring the Mathematics: Interactive Tools

The use of technology in mathematics has been found to contribute to student conceptual understanding by allowing for meaningful exploration particularly in association with the relationships between mathematical ideas (National Council of Teachers of Mathematics [NCTM], 2011). To that end, students can further explore the Platonic Solids using *Interactives: 3-D Shapes* (learner.org). The previously discussed SFA combined with this Interactive, offers students an opportunity to compare the five Platonic Solids. Included in the *Interactive* is practice manipulating each of the five Platonic Solids to determine their defining characteristics. This is followed by a 39-question multiple-choice assessment that can be printed as evidence of understanding. Through exploration with the interactive online tools, students should discover that there are only five Platonic Solids and that Euler's Formula states that for any polyhedron that does not intersect itself, the number of faces plus the number of vertices minus the number of edges will always equal two ($F + V - E = 2$). Students can use this formula and their explorations to complete a chart of what they are observing (see figure 5.4).

Name	Image	Face Shape	Number of Faces	Number of Edges	Number of Vertices
Tetrahedron		Triangle	4	6	4
Cube	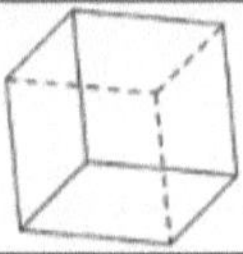	Square	6	12	8
Octahedron	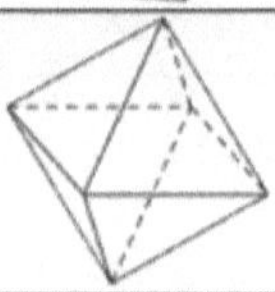	Triangle	8	12	6
Dodecahedron	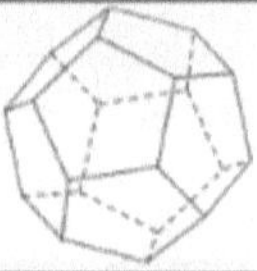	Pentagon	12	30	20
Icosahedron	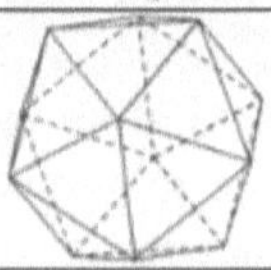	Triangle	20	30	12

Figure 5.4 **Types and Characteristics of the Platonic Solids.** *Source*: Created by authors

By using the activities presented—previewing the text and vocabulary, and exploring the mathematics with a SFA and through interactive tools—students will be better equipped to understand the mathematical concepts that they encounter when they engage in a close reading of *All of the Above*. Providing pre-reading strategies that laid a solid foundation for students' comprehension was important. These activities also provide a framework for students to revisit as they complete the while- and after-reading activities.

WHILE READING *ALL OF THE ABOVE*

While-reading strategies are used to remind students to engage with the text as they monitor their comprehension of their reading. For *All of the Above*, character and setting play a significant role in shaping the plot. The following section outlines and describes strategies and activities that English Language

Arts (ELA) and mathematics teachers can use while students are reading *All of the Above*.

Making Predictions

Comprehension is often strengthened when students make predictions about texts they read. Students who make predictions use information read in the text, in addition to their prior knowledge, to predict what might happen next in the plot. Secondary students need opportunities to justify the reasoning behind their thinking. Asking students to provide rationales (either orally or in writing) allows for metacognition, or thinking about their own thinking. When teaching students how to make predictions, we must emphasize the importance of monitoring and adjusting our predictions as we gain more information through reading. To implement this strategy with your students, prepare a series of questions from *All of the Above* that warrant opportunities for students to stop and think about how the setting may be impacting the plot or how characters are responding to various events within the story. Possible questions might include the following: *Who do you think destroyed the tetrahedron project and why (p. 149)? What do you think the motivation behind the vandalism was? What do you think will happen next? Do you think Rhondell will rejoin the math group or avoid negative feelings she has suppressed (p. 171)? How do you think the author, Shelley Pearsall, feels about Rhondell (pp. 227–230)?* Provide students with time during reading to respond to your questions and reflect or revise as necessary in their reading journals, using textual evidence to support their responses.

Setting

Time, historical context, location, weather, and social conditions all comprise elements of setting. Talking about setting helps readers better understand the characters and their reactions to major events in stories, including conflict and resolution. For *All of the Above*, ask students to think about how the story's plot might have changed if the setting had changed. Upon completion of the text, teachers may facilitate a discussion with the class about how the setting impacted the characters and the plot with questions such as the following: (1) *If the tetrahedron had been created in a private school, how might the catastrophe of the giant tetrahedron being destroyed (p. 111) have been prevented? (2) Would the tetrahedron have been destroyed if the math club took place in a rural school? Why do you think that?* Ask students to consider the setting of *All of the Above* as they read independently. Provide students with 10 sticky notes each. Ask them to list five different settings that they encounter in the text. Then, on the other five sticky notes, ask students to write about how they think each setting impacts the plot. As students track the setting and impact on the plot, ask them to place their sticky notes on a T-Chart in the classroom. The setting T-Chart should include a column for the various settings identified by the students and a column for noting impacts in plot. The T-Chart can be developed gradually, over time, as students read the novel. Having it readily available to view on a continual basis, students' thinking and comprehension about the story line and the characters as they read can be realized.

Characterization

As students read, they may have difficulty remembering specific details about each of the main characters: Rhondell, Marcel, James, and Sharice. One way to deepen student involvement with ever-evolving characters is by implementing the open-mind portrait strategy (see figure 5.5 for an example of an open-mind portrait for Marcel). When students create open-mind portraits, they create mental images leading to deeper comprehension of the text since visualization helps students engage deeply with the characters they read about (Joyce, 2018). Begin by giving each student two pieces of paper for each open-mind portrait they will create. Have students draw and color the neck and head outline for the character they are studying. Staple a second sheet of paper behind the drawing. Cut out the sheets of paper in the shape of the drawn head/neck outline. On the top sheet of paper, have students draw what

 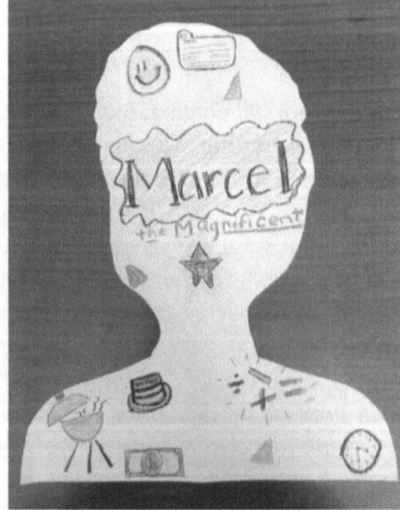

Figure 5.5   Open-Mind Portrait for Marcel. *Source*: Created by authors

they think the character looks like. On the inside of the open-mind portrait, have students add what they think would be inside the character's mind based on what they are reading in the text. As they continue to read and learn more about the characters, they may continue adding sketches, words, phrases, or quotes that best represent the character.

Comprehension and Mathematical Discussion Guides

Explicit comprehension instruction helps ensure students understand what they read. The discussion guides that follow serve as examples of the types of questions teachers may ask students. Table 5.1 provides a list of questions representing different comprehension types, helping students better understand story elements.

Likewise, it is important for students to discuss the mathematical concepts found in *All of the Above*. Topics include geometric solids with an emphasis on tetrahedrons, algebraic reasoning, scaling, conversion, and more. Table 5.2 includes examples of questions teachers can use with students during the

Table 5.1 *All of the Above* Comprehension Discussion Guide

Topic	Questions
Summarize Important Ideas and Supportive Details	Read the pages aloud to your students. Ask students to identify the main points. Ask them to identify details that support those ideas. Ask students to use words from the details to summarize the text in 20 words, including only the most important points from the pages. (pp. 49–55)
Make Connections with Important Ideas in the Text	Describe a time when you had to re-start a project after completing it. How did that make you feel? (pp. 113–114)
Integrate New Ideas with Existing Background Knowledge	Rather than telling James why she doesn't want to be in the math group, Rhondell says she is "going to be late for class" and "can't talk any longer". Why do you think she kept the truth from James? (p. 172)
Sequence of Events	Predict if Rhondell will rejoin the math group or avoid negative feelings she has suppressed. (p.171)
Responses to the Text	IN what ways if rebuilding the tetrahedron worth it? Who do you think might rebuild it, if so? (p. 149)
Visualize Characters, Settings, & Events	Read students the passage. Pause. Ask students to jot down what they visualize. They may write words or phrases directly from the read aloud or they may write down other words they are thinking of as the passage is being read aloud. They may create sketches of the images they imagine. Ask several students to orally share their mental images. Discuss how their interpretations might be different based on their own experiences and backgrounds. Remind students to regularly create mental images to help them better understand parts of the text that include rich description. (pp.127–131)

Table 5.2 *All of the Above* Mathematical Topics Discussion Guide

Question	Response
The text tells us that tetrahedrons are geometric solids with four faces, and that all four faces are equilateral triangles. What could you add to provide more detail about the characteristics of tetrahedrons?	Tetrahedrons are a special type of geometric solid known as Platonic Solids. Tetrahedrons are composed of four equilateral triangles as the faces with none of the faces being parallel. There are a total of six edges and four vertices on each tetrahedrons. Each vertex is created by the connection of three faces (p. 3).
The text states that the largest tetrahedron ever constructed was approximately 7-feet tall, and it was made of 4,096 smaller tetrahedrons. How could you determine the length of time that it would take to create this large tetrahedron?	First, determine how long it takes to create one tetrahedron. Next, determine how long assembly of multiple tetrahedrons takes. Since it is known that a total of 4,096 smaller were used for the larger tetrahedron, the amount of time it takes to assemble a multiple of two could be used. For instance, if 32 smaller tetrahedrons were assembled in 90 minutes, you could multiply 90 times 128 (since 32 × 128 = 4,096) to get an estimated time for the total build of the large tetrahedron (p. 3).
Mr. Collins and his seven students in the math club made a goal of 16,384 smaller tetrahedrons to build a giant tetrahedron. They knew that each person could make about 30 smaller tetrahedrons in an hour for five days per week. What are other factors that might influence the calculation of finding out how long it would take to assemble the giant tetrahedron?	Putting together the smaller tetrahedrons can be quite tricky. What if the assembly collapses during the build? What if someone accidentally trips and falls on the structure. Do all weeks have five school days? What about school holidays or if someone in the group is absent (p. 39)?
Willie Q's Cannonball Cornbread recipe calls for twelve unique ingredients. To scale the recipe up or down, you multiply or divide by the number of people served. Which of the ingredients in this recipe may be the most complex to scale up or down? Explain.	This recipe calls for a dash of nutmeg and one egg. Both of these units are nonstandard measurements, thus making conversion more complex. The recognized conversion of one egg is one-fourth cup of egg substitute, therefore the more standard unit of cup can be used. A dash is approximately one-eighth of a teaspoon; therefore the more standard unit of a teaspoon can be used (p. 74).

Table 5.2 (Continued)

Question	Response
If the Tangy Tetrahedron Barbeque Sauce recipe created by Willie Q was enough for 12 sandwiches, how could you determine the quantity needed for each ingredient to feed a crowd of 144 people? What measurements might need converting into other units?	By multiplying each value by twelve, we can determine the quantity needed for each ingredient: 24 tablespoons butter, 36 tablespoons onion, 6 cups celery, 12 tablespoons brown sugar, 12 tablespoons Worcestershire sauce, 24 tablespoons malt vinegar, 3 cups lemon juice, 12 teaspoons dry mustard, and 12 cups ketchup. It would make sense to convert the tablespoons to cups for ease of measuring. There are 16 tablespoons in a cup, so by dividing the number of tablespoons by 16 you can determine the number of cups needed for each ingredient. Similarly, teaspoons can be converted to tablespoons. Three teaspoons is the same as one tablespoon, so by dividing the number of teaspoons by three you can determine the number of tablespoons needed for each ingredient (p. 220).
Due to the repeating pattern found in tetrahedrons, the structure can be extended to infinity. How can you determine the next number of smaller tetrahedrons needed to create an even larger tetrahedron than the one created by Mr. Collins and his students?	We can see a pattern that with each new level used to create a tetrahedrons, you must multiply by four. It takes four small tetrahedrons to build a level one structure. Level two takes 16 small tetrahedrons. This pattern continues with 64, 256, 1,024, 4,096, 16,384, and so on. Therefore, the next largest tetrahedron would need 65,536 small tetrahedrons (16,384 × 4) (p. 235).

reading of the text to prompt mathematical discussion related to the embedded mathematical concepts.

AFTER READING *ALL OF THE ABOVE*

After-reading strategies are primarily used to invite students to respond to the text. Here we describe activities—Carousel Brainstorming, Creating Three-Dimensional Shapes, and Tetrahedrons to Infinity—that offer students opportunities to think and talk about what they have read in *All of the Above*.

Carousel Brainstorming

Carousel brainstorming (Simon, n.d.) may be another impactful way to discuss the outcomes of the text, as it will allow for opportunities to reflect upon the book by engaging in dialogue with small groups of peers. Ask groups of three to four students to respond to prompts displayed around the room on chart paper. Student groups move around the classroom answering each posed question or statement. For accountability and group management, provide each group with a different color marker to use when answering the prompts. This way, the teacher can track group contributions and follow up with individual teams as necessary. When a group arrives at a new prompt, they may read what previous groups wrote and add to or respond to previous comments. Prompts to include for the Carousel Brainstorming may include the following:

- James immediately thought of objects from the real world when presented with three-dimensional figures in Mr. Collin's class (p. 7). Where can tetrahedrons be found in the real world?
- Marcel and Willie Q run a restaurant and must consider the amount of food to prepare daily in order to make profit. Marcel states, "Nobody eats barbecue in January" (p. 91). If Marcel and Willie Q make 24 batches of Marcel's Slow Burn Sauce (p. 48) in June, and sales are down to 50% in January, select one of the ingredients and describe how much they will need to make the sauce in January? Use mathematics to support your description.
- Throughout the text, Mr. Collins shares facts about tetrahedrons (pp. 3, 99, 199, and 235). How do the characters in the story mirror the qualities of a tetrahedron?
- In what ways is each character (Rhondell, Marcel, James, and Sharice) important to the project?

Creating Three-Dimensional Shapes

In order to reinforce the mathematical understanding of three-dimensional shapes, including tetrahedrons, students can work together to create various three-dimensional shapes using nets—2D versions of 3D figures (Altes, 2020). Students can then use these models to calculate the surface area of each shape. Using printable nets for the Platonic Solids, such as the net for tetrahedrons found on page 248 of the text, students can cut, fold, and assemble the nets into three-dimensional figures. These can stimulate classroom discussions about the characteristics of each figure. Prompts to stimulate the conversation may include questions such as the following: *What do you notice or wonder about each figure? How many edges does the octahedron have?*

A great way to connect directly with the text would be to attempt to construct the same giant tetrahedron as the one in *All of the Above*. Students could complete this task as a class. One suggestion to account for the amount of time needed for this project would be to ask students to pledge to create a specific number of individual smaller tetrahedrons on their own and then bring them to class for assembly. Many mathematical connections could be made with questions such as the following: *How could we find the surface area of one individual smaller tetrahedron or for the entire giant tetrahedron? How long will construction of the giant tetrahedron take based on the rate of the smaller tetrahedron creation? What is the ratio of one smaller tetrahedron to the midsized tetrahedron and to the giant tetrahedron?* The completed tetrahedrons can be displayed for parents or the community at a school open house or family night.

Tetrahedrons to Infinity!

In the novel, Mr. Collins describes how tetrahedrons, because of their repeating patterns, can expand to infinity (p. 235). To explore this concept of infinity, students can create tetrahedrons using materials such as toothpicks for the edges and either marshmallows or gumdrops as the vertices. If using marshmallows, be sure to let them sit out for at least 24 hours before the activity; this makes them easier to work with and less sticky. To assemble one tetrahedron, students will use six total toothpicks and four gumdrops. Three toothpicks in the shape of a triangle will serve as the base. The toothpicks will meet and connect to three gumdrops which represent vertices. The remaining three toothpicks will connect at the existing gumdrops (vertices) and meet at a gumdrop on top (see figure 5.6).

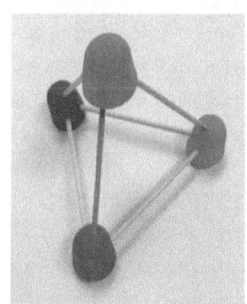

Figure 5.6 Tetrahedrons to Infinity—Base, Initial Tetrahedrons, and Level One Tetrahedrons. *Source*: Created by authors

Ask students to think about how they could combine these individual tetrahedrons to demonstrate a repeating pattern that can expand to infinity. They need to consider that once combined, they will have to remove some gumdrops as the vertices are shared between the newly combined figures. The connection with infinity will be seen as the number of tetrahedrons needed to make each new level will always increase by a multiple of four (1, 4, 16, 64, 256, and so on).

BEYOND *ALL OF THE ABOVE*

Integrating multiple subject areas in teaching and learning has many benefits for students. According to Repko (2008), "Interdisciplinary learning promotes higher order cognitive abilities" (p. 176). Students who learn through an interdisciplinary approach are also able to recognize bias more easily, think more critically, learn to embrace ambiguity, and consider ethical issues (Field et al., 1994). *All of the Above* naturally lends itself to integrated study with multiple subject areas. With these benefits in mind, the following extension activities include integration with art, science, and social studies as connected to the text.

Writing an Epilogue

After reading novels with strong characters such as those found in *All of the Above*, readers might ask what happened to the characters after the story ended. One way teachers could help students resolve these "I wonder" questions would be to guide students in the writing of an epilogue for *All of the Above*. Prompt students to select a character they studied throughout *All of the Above*. This might even be the character studied using the open-mind portrait. Teachers can ask students to think about how their character evolved throughout the story, revisit difficulties their character faced, and review problems they solved. Ask students to imagine what life might be like for each of the characters in six years (after graduating from high school). Students should take into consideration what they know about the evolution of the character as they write their epilogues. Instruct students to write a one-page epilogue that describes how life evolves for the character, detailing realistic problems the character may continue to face after graduating from high school. After students have written the epilogues for their selected characters, group them according to the characters chosen. Have each group discuss their epilogues comparing the various perspectives and textual evidence provided. Lastly, have each group share out during a whole-class discussion.

Recipes

Throughout the novel, the author includes multiple recipes related to the characters and the events in the story. The class can work together to create a recipe book with their own recipes or from recipes shown in the book, such as *Willie Q's Chocolate Truth Cake* (p. 137). In class, they will apply their understanding of ratio reasoning to convert measurement units for each recipe to accommodate 4 people, 12 people, and 20 people, or other agreed-upon quantities. Like the class party described on pages 69 to 90 in the text, students may bring in their recipe creations for the entire class to enjoy!

Three-Dimensional Book Report

For this extension, teachers can provide students with the opportunity to each create their own three-dimensional book report. To begin, provide each student with 16 copies of the tetrahedron template with tabs from the novel (p. 248). On at least 10 of the tetrahedrons, ask students to include the following:

- Title of novel: Ask students to write the title of the novel on one of the tetrahedrons.
- Illustration: Ask students to create illustrations from students' favorite scenes from *All of the Above*.
- Characters: Ask students to think about how they visualize the characters from the novel. Ask students to select their favorite character and create a sketch of him/her. Encourage students to utilize textual evidence when illustrating their characters.
- Impactful Quotes: Ask students to think about which quotes from the novel with which they had the greatest connection. Ask students to write the quote and corresponding page number on one of their tetrahedrons.
- Theme of the Story: Ask students to reflect on the plot of the story, and then ask them to consider their key learning outcome. Ask students to use words and illustrations to represent the theme, message, moral, or motif from *All of the Above*.
- Setting: Ask students to sketch a picture of Mr. Collins's high school classroom on one of the tetrahedrons.
- Conflict/Problem: Ask students to consider the types of conflicts in the novel. Were the conflicts Man versus Man, Man versus Society, Man versus Self, or Man versus Nature? Ask students to write their responses on paper and then add to their three-dimensional book reports.
- Resolution: Ask students to write the resolution of the story on another tetrahedron.

- Opinion: Ask students to think about their opinion of the novel. What adjectives would they use to describe *All of the Above*? Have students create an adjective word cloud on a post-it note that represents their feelings about the book.

After students have added all of the required information for each of their tetrahedrons, ask them to glue their 16 tetrahedrons together to create a pyramid (see figure 5.7)—their own three-dimensional book report.

After their pyramids have been constructed, students may embellish their 3D book reports to showcase their knowledge and creativity using artifacts that showcase their understanding of the various story elements from *All of the Above*.

Art

Art can be a strong motivator as students work on learning about geometric concepts (Brezovnik, 2015). The main geometric concept in *All of the Above* is three-dimensional figures. Many famous works of art lend themselves well to connecting to geometry and specifically three-dimensional shapes. Pablo Picasso is well-known for his work with cubism. In this style of artwork, simple geometric shapes are used on a two-dimensional surface to represent three-dimensional shapes. After viewing an example of one of Picasso's cubism masterpieces, such as *Three Musicians*, students could take a photograph

Figure 5.7 **Tetrahedron Book Report Example.** *Source*: Created by authors

of an everyday object and then replicate it using cubism techniques. M. C. Escher is another influential artist famous for his use of geometry in art. In *Double Planetoid*, Escher combines two interlocking regular tetrahedrons. Students could use this image as an inspiration to create other three-dimensional figures using modular origami. Similar to regular origami, modular origami uses paper folding without any types of adhesives to create three-dimensional figures. The unique aspect of modular origami is that it is done with multiple interlocking pieces of paper rather than the one piece used in traditional origami designs.

Science

Science standards for middle-level students often include a life science performance expectation about the dynamic relationships within an ecosystem. The characters in *All of the Above* live in an inner-city setting among family, friends, and acquaintances that share a common bond. The community comes together to support James, Rhondell, Sharice, Marcel, and Mr. Collins to reach their goal of building the world's largest tetrahedron. The concept of interdependence among the living things in an ecosystem is akin to the community in Alexander Hamilton Middle School in *All of the Above*. Related to the science of interdependence in an ecosystem and with collaboration among a community or team, students can benefit from participating in an activity called "Can you build it?" Prior to the activity, the teacher should create five different unique constructions from everyday objects. This can be anything from paper plates to craft beads to building blocks (see figure 5.8).

The construction should be simple enough to be recreated in less than two minutes but complex enough to be a challenge. Each of five unique

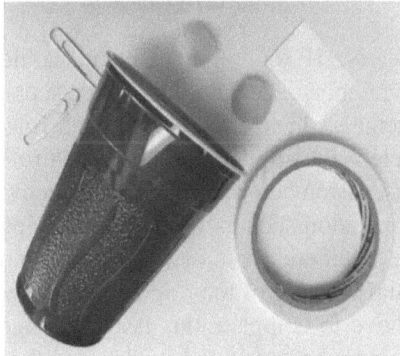 

Figure 5.8 Can You Build It? Example Materials and Sculpture. *Source*: Created by authors

structures will be used in separate rounds of the activity. To begin the activity, place students in teams of four. Assign each student a role for the team: Eyes, Hands, Ears, and Brain. The Hands is the only team member that gets to actually see the original structure. This person will get to look at the structure for exactly one minute. During this observation, there is no talking, touching, or writing. They will then return to the team and explain the unique creation. During the explanation, Eyes cannot use any hand motions! It is best to either place your hands in your pockets or behind your back. The Hands is the only team member that gets to actually touch the construction materials. This person will listen carefully as Eyes describes the creation and will try to replicate it exactly. This person is not permitted to speak during the construction. The Ears will listen to the Eyes describe the construction and also watch Hands carefully build the construction. This person is responsible for asking questions to clarify the information being shared. They must be attuned to communication signals among all team members. The Brain supervises the other team members to make sure they are following the guidelines of their roles. They are not to interject in anyway other than to remind members of their responsibilities. All teams are given a tray or baggie with identical materials used in creating the original structures. For an added challenge, extra materials can be included. For the first round, all Eyes are invited to a platform outside of viewing range of the rest of the class (a large box works well). The Eyes are given one minute to look at the structure. They may walk around to get different perspectives but may not touch it, talk about it, or take notes. After one minute, the Eyes return to their team. Another timer is set for two minutes as the Eyes describe the structure and the Hands build. After the two minutes are up, show the original structure to the class and give time for them to compare to their newly built structures. Ask questions to the group about what they experienced, such as the following: *What was the most difficult part of the task? What might have made it easier?*

Rotate the roles among the team members. All four team members should get a chance to experience each role. Repeat this process until there is just one original structure left. After all team members have experienced each role, have them come together as a team to select a team member to fill each role for one final round. It is likely that the discussion between rounds has uncovered the need for more forms of communication such as writing or drawing to clarify ideas. For this round, Eyes are given one piece of white paper and one pencil to take notes during their initial viewing of the original structure. Repeat the process as for the rounds above to complete this last challenge round. Lead one final class discussion about what students have learned during this activity. Connect back to the interdependence among organisms in an ecosystem and collaboration among team members. Highlight the need for unique talents and effective communication, whether that is body language, verbal, or written.

Social Studies

ELA standards often encourage a connection to history and social studies. Marcel's father, Willie Q, owned a barbecue restaurant and was a veteran of the Vietnam War. Mr. Collins's brother fought and was killed in the Vietnam War. To learn more about this period in our history, and to understand the importance of using primary sources, students can interview a Vietnam veteran either independently or as a class through a virtual interview. After the interview, students should write a report summarizing the information obtained during the interview. Along with that, students can relate this information to their own lives and with the lives of the characters in *All of the Above*. If students conducted separate interviews, they could also present their reports to the class to find commonalities between the stories.

CONCLUSION

All of the Above is a compelling story of a teacher and four of his students as they develop mathematical competence through an after-school math club. Readers can connect to the students' struggles and triumph as they build the world's largest tetrahedron in ways that are personal, relevant, and meaningful. This novel is a wonderful addition to the growing body of mathematics-themed literature written for young adults. As secondary students read widely and for a variety of purposes, providing relatable literature that supports mathematics instruction is important. The activities shared in this chapter related to *All of the Above* provide students with opportunities to become both lifelong readers and mathematically minded citizens.

REFERENCES

Altes, G. K. (2020, April 29). *Paper models of polyhedra*. https://www.polyhedra.net/en/.

Brezovnik, A. (2015). The benefits of fine art integration into mathematics in primary school. *CEPS Journal, 5*(3), 11–32.

Dweck, C. S. (2008). *Mindset: The new psychology of success*. Random House Digital, Inc.

Field, M., Lee, R., & Field, M. L. (1994). Assessing interdisciplinary learning. *New Directions in Teaching and Learning, 58*, 69–84.

Gunning, T. G. (2020). *Creating literacy instruction for all students*. Pearson.

Joyce, B. (2018). *Reading: Learning about open minded portraits*. Open Mind Portraits. https://ourglobalclassroom.org/2018/03/19/reading-learning-about-open-minded-portraits/.

National Council of Teachers of Mathematics (NCTM). (2011). *Strategic use of technology in teaching and learning mathematics [Position statement]*. https://www.nctm.org/Standards-and-Positions/Position-Statements/Strategic-Use-of-Technology-in-Teaching-and-Learning-Mathematics/.

Repko, A. F. (2008). Assessing interdisciplinary learning outcomes. *Academic Exchange Quarterly, 12*(3), 171–178.

Simon, C. A. (n.d.). *Brainstorming and reviewing using the Carousel strategy*. ReadWriteThink. http://www.readwritethink.org/professional-development/strategy-guides/brainstorming-reviewing-using-carousel-30630.html.

Tompkins, G. E. (2004). *Teaching vocabulary: 50 Creative strategies, Grades K–12*. Pearson.

Chapter 6

Reasoning with Equations and Functions in *The 5th Wave*

Rebecca Grice Gault and Jennifer Edelman

The 5th Wave (Yancey, 2013) is a young adult science fiction novel about the survival of teens forced to fight for the salvation of humankind following an alien invasion. The major literary theme of the novel—being human in an increasingly inhumane world—unfolds in unexpected ways as the teens learn that those they trust may not be who they seem. Because the protagonists in *The 5th Wave* encounter crises that require critical problem-solving, many connections to concepts across secondary mathematics courses are possible. In this chapter, we present an eight-week mathematics and language arts book study that covers topics typically learned during an algebra course. The mathematics topics that are explored include creating equations, reasoning with equations and inequalities, interpreting functions, and building functions. Our literacy focus is the study of dystopia as a genre and includes the development of critical lenses through which we examine literature and, by extension, culture.

THE 5TH WAVE BY RICK YANCY

The 5th Wave (Yancey, 2013) follows the experiences of Cassie, Ben, and Evan as they try to make sense of an alien invasion and find ways to save those they love and all that is left of humankind while attempting their own survival. As the readers meet the first narrator, Cassie, they learn that the aliens have already attacked Earth with four waves: technological destruction, earthquakes, viral pandemic, and invasion by aliens who appear to be human. Cassie is on a quest to find her brother, Sam, as the fifth wave begins. She finds herself allied with Evan and Ben, teens she knew before the invasion but is not sure she knows now. Cassie must decide who to trust as she faces the possibility of human extinction and a fate worse than death for herself and those she loves—the loss of what it means to be human in the face of fighting the inhuman.

CONNECTING MATHEMATICS AND LITERACY

Literacy instruction and mathematics instruction have a shared goal of incorporating critical thinking into students' learning, often referred to as "making sense" in both disciplines (Brozo & Crain, 2018; Draper, 2002). The National Council of Teachers of Mathematics (2014) calls for integration of discussion, argumentation, and written explanation into the mathematics classroom. The incorporation of studies of novels such as *The 5th Wave* has strong potential to engage students in critical sensemaking across literacy and mathematics content standards. Furthermore, literacy can play a crucial

Table 6.1 Overview of Novel Reading Schedule with Mathematics Teaching Focus

Week	*The 5th Wave* Reading	Plot Point Mathematics Connection	Concurrent Mathematics Teaching Focus
1	Prologue Chapter 1: pp. 1–10	Cassie is alone and must manage her supplies of food and other goods while searching for more food and replacement items.	Creating and solving linear equations and inequalities in one variable.
2	Chapter 1: pp. 11–24	Cassie learns that the world population is decreasing due to successive "waves" of alien interference. Cassie notices alien drones arriving in the sky each day.	Creating, solving, and graphing linear equations in two variables.
3	Chapter 2 Chapter 3	Ben is informed by Dr. Pam that "one out of every three surviving human beings on Earth is one of them."	Creating, solving, and graphing linear equations and inequalities in two variables.
4	Chapter 4 Chapter 5	Ben becomes aware that the human population is decreasing while the alien population may be increasing.	Representing, constructing, and solving systems of equations.
5	Chapter 6	Vosch says they are in a "war of attrition" and the question is how long they can last with the supplies they have.	Rearrange formulas to highlight a quantity of interest.
6	Chapter 7 Chapter 8	Ringer tells Ben "we're the 5th wave" which brings up the question: how many humans have been turned into soldiers?	Function concepts and function notation.
7	Chapter 9	Evan asks Cassie what she would do if the Earth was dying.	Interpreting functions by table, graph, and equation. Graphing functions. Comparing functions in different representations.
8	Chapters 10, 11 12, 13	Ben and Cassie meet again and find that they both know the same secret.	Writing arithmetic sequences and geometric sequences.

role in supporting mathematics education that may not always be recognized at the surface level. Teachers can use problem-solving situations from novels to engage students in curricula that promote deeper understandings of mathematics concepts, transferring mathematical knowledge to new contexts and representations, and learning to make meaning of topics that intersect with real-world problems (Draper, 2002). Mathematics teachers often report challenges to incorporating literacy into mathematics teaching and learning such as time constraints, seeing the elements of reading instruction as another teacher's role, and finding literacy strategies not relevant to mathematics reasoning (Brozo & Crain, 2018; Donahue, 2003). However, deep learning of mathematics concepts depends on the willingness of students to engage in practices of making sense and applying concepts to contexts, and these are practices that align closely with literacy instruction (Donahue, 2003). For these reasons, we propose the pairing of the first algebra course with a book study of *The 5th Wave*.

We have included the reading schedule (see Table 6.1) for the book as a preview to the before-, during-, and after-reading activities, along with plot points that connect to mathematics topics, and the concurrent mathematics teaching focus for each plot point.

BEFORE READING *THE 5TH WAVE*

With an overall focus on developing students' understanding of the concept of criticality, we begin with a definition. According to Gholdy Muhammad (2020), criticality "is the capacity to read, write, and think in ways of understanding power, privilege, social justice, and oppression" (p. 120). The story contained in *The 5th Wave* leads the reader to question how and why different people, cultural norms, locations, and habits survive as humanity itself is being destroyed. When viewed through the lens of criticality, students can begin to see how the intersection of privilege, social justice, and oppression work to either save or destroy the world as we know it.

Defining Dystopia

Eighth- and ninth-grade students in the United States have likely had the opportunity to engage with a number of dystopian books such as *The Hunger Games* (Collins, 2008), *1984* (Orwell, 1949), and *The Giver* (Lowry, 1993). They may have engaged with the genre through movie versions of *The Hunger Games* trilogy (Ross & Lawrence, 2012, 2013, 2014, 2015) or *War of the Worlds* (Spielberg, 2005) and television shows such as *The Handmaid's Tale* (Miller, 2017–2021) or *The Walking Dead* (Darabont &

Kang, 2010–2021). Prior to reading the novel, it is important to activate their prior knowledge of the dystopian genre. We recommend using a survey to determine which dystopian stories students are most familiar with and then organizing a compare/contrast session across stories, in pairs or small groups. Whatever method you chose, the goal is to come up with a list of criteria that identify a story as a dystopia. These criteria may include things such as existing in a futuristic world; an event that upsets the entire structure of a society; the rise of a totalitarian government, society, or business; a hero or heroine who overcomes loss to save the world; and a message that could be applied to our current circumstances (the "it's not too late if only you would change your ways now" message). Once you have determined your criteria for identifying a dystopia, apply the list to the movie *WALL-E* (Stanton, 2008), which many students have probably seen. If not, you could arrange a showing in your classroom. Students may be surprised to discover that despite its upbeat appearance, this movie actually tells a dystopian tale.

WHILE READING *THE 5TH WAVE*

Layered Texts

The events in *The 5th Wave* relate to the modern world, making it easier for students to "see" themselves within the story. To develop students' ability to view the story and world using a lens of criticality, while simultaneously developing their knowledge of the dystopian genre and mathematics concepts, pairing the weekly book readings with nonfiction texts that address similar topics can be beneficial. Layering texts happens when two or more connected texts are used to examine a topic or viewpoint and "help students understand local, national, and global communities . . ." (Muhammad, 2020 p. 147). Topics that align with the action in *The 5th Wave* as a starting point for educators are suggested. Please note that text includes any consumable media such as videos, songs, podcasts, artwork, posters, political cartoons, and the like. Some general ideas for using layered texts are offered, though there are many excellent resources for pairing fiction and nonfiction available online such as ReadWriteThink.org. and AchievetheCore.org.

While not an exhaustive list, here are some topics that can be used in concert with *The 5th Wave*: the search for life on other planets, traditions for naming children around the world, food deserts and food insecurity, gun violence and children, cultural resources saved during times of conflict, pandemics (sources, medical care, etc.), drone warfare and surveillance, technological implants (biohacking), population growth, climate change, internment camps, parasites and scavengers, chocolate production, separating children from families during immigration, women in the military, child soldiers, the development of "women and children first" policies, colonization, psychology of lying, working undercover, evolution, the use of facial recognition in policing, and Stockholm syndrome.

Fiction versus Nonfiction

When layering texts, it is important to distinguish between fiction and nonfiction. Often students have a simplistic view of nonfiction as "true" and fiction as "made-up." By comparing and contrasting the topic of the nonfiction piece with the chapters read in *The 5th Wave*, we can help students see how the line between fiction and nonfiction is sometimes blurred. For example, pairing Cassie's search for water with the text, "The Navajo Nation Faced Water Shortages for Generations—and Then The Pandemic Hit" (Calma, 2020), or using the article, "The World's Population Is Nearing 8 Billion, That's Not Great News" (Davis & Griggs, 2019), when exploring the ratios of humans to aliens in Week 4 of the reading would provide such opportunities. Teachers may ask students to do this independently, in small groups, or as a whole-group activity. This can also lead to a discussion of an author's craft as you compare how Yancy presents the same topic as the author of the nonfiction text.

Podcast (or Social Media) Updates

In this activity, ask students to imagine that there is a resistance movement during *The 5th Wave* that communicates important news via podcasts (or other social media tools). Students participating in this activity will be responsible for producing a three-to-five-minute podcast (or other social media product) for the characters in the section you are currently reading. As an example, you might consider assigning a podcast for Zombie's group (described in Chapter 43) and the text, "The True Story of Brainwashing and How It Shaped America" (Boissoneault, 2017). Students should use information from the nonfiction text and should address a problem that the characters are currently experiencing.

Debate

Debate is a staple in many ELA classrooms and it can be an effective tool for critically examining text. We recommend engaging students in the identification of viewpoints in the chapter and/or nonfiction text, and then assigning a group to each. For example, you might have students assume the roles of Cassie and Crisco and debate the ethics of scavenging for valuables as described in Chapter 12 (pp. 69–70). Pairing the article "If You Are Old

Enough to Carry a Gun" (Burke & Hatcher-Moore, 2017) with the action in Chapter 56 of *The 5th Wave* would provide the opportunity for students to debate the topic of child soldiers. You may also consider having students assume a role such as a character from the book, the author of a nonfiction text, or the subject of a nonfiction text.

There are myriad resources available to assist in setting up debates in a class. First, it is important to involve the entire class, so teachers may want to put students into small groups. In these groups, teachers can assign one student to give an opening statement, another to prepare the main arguments for the team, a student in charge of preparing a rebuttal statement which includes anticipating the other teams' arguments, and a student who will give the closing argument. Identify three to five students as judges who must weigh the effectiveness of the arguments. As students prepare to debate, the use of a Persuasion Map to help organize their arguments may be beneficial.

Mathematics Connections Activities

Because of the many problems the characters in *The 5th Wave* encounter in the plot of the novel—some intersecting with issues of science, others with issues of social justice, and still others focused on practical issues of supply and economics—almost any plot point in the novel can be used to generate a mathematics problem. We have developed an overview of a unit focused on topics from a first course of algebra, but plot points could also be used to develop units or lessons focused on topics covered in middle-grades mathematics, geometry, or the second course of algebra as well.

In this section, we present the activities as they occur each week. Each activity aligns with assigned reading for the week. While an example of a context-related task is given for each week, additional context-related tasks could be developed for each lesson throughout the week. Teachers might choose to organize lessons for the week such that one day is chosen as "book study" day when class work is focused on solving problems relevant to the struggles and obstacles encountered by Cassie, Ben, and Evan, or teachers might choose to incorporate tasks related to contexts from the novel throughout the week.

Week 1: Linear Equations and Inequalities in One Variable

Students are reading about Cassie's belief that she is one of the few survivors on Earth and her constant struggle to find food and supplies. Have students create and solve this inequality in one variable related to Cassie's supplies:

Cassie has 5 bottles of water remaining. She must drink at least 17 bottles of water over the next week to avoid dehydration. If Cassie knows she can carry only 6 bottles of water each time she finds a store, at least how many trips will Cassie need to make to avoid dehydration?

Students should create the inequality $6x + 5 \geq 17$ and solve for x to find that Cassie will need to make at least two trips to obtain the required number of water bottles. Students may need help understanding that the words "at least" in the problem statement correspond to creating an inequality rather than an equation. Students may also need help understanding the relationship between "at least" and the greater-than-or-equal-to symbol.

Week 2: Linear Equations in Two Variables

Cassie recounts her experiences with the previous waves, the death of her mother and father, and the taking of her brother at a human camp. The students learn that the world population is decreasing due to successive "waves" of alien interference and that Cassie has noticed alien drones arriving in the sky each day. This week's reading provides an opportunity to explore interpreting graphs of linear equations in two variables. Provide students with the two graphs shown in figure 6.1 and begin a discussion with the following questions: *What do you notice?* and *What do you wonder?* Ask students how these two graphs represent Cassie's observations on alien drones and reflections on the human Earth population. Students should be able to identify which graph has a negative growth rate (negative slope) and therefore matches Cassie's reflections on the human population, and which graph has a positive growth rate (positive slope) matching Cassie's observations of alien drones. Please note that we have developed an interpretation of Cassie's narration that shows two drones being added each day, but this could be adapted.

Once the context of each graph is established, have students write a title for each graph and label the axes. The novel makes it clear it would be reasonable to label the axes of the drone graph as days for the independent variable (x-axis) and number of drones for the dependent variable (y-axis), but the

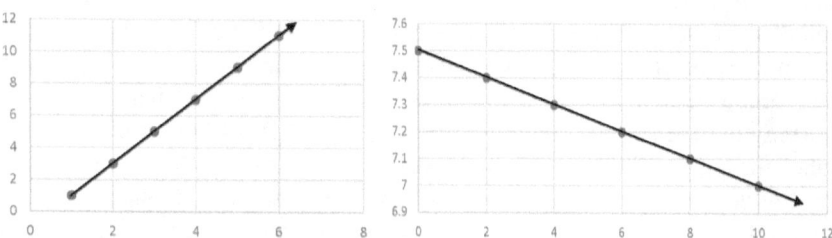

Figure 6.1 **Graphs to Use in Week 2.** *Source*: Created by authors

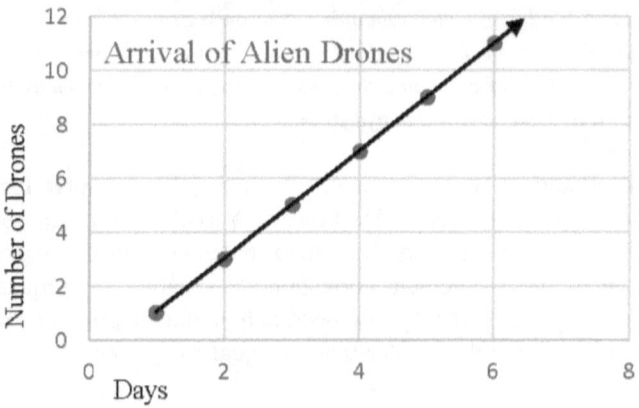

Figure 6.2 **Labeling of Drones.** *Source*: Created by authors

number of drones being added to the sky each day is not clear so this is left for the teacher or students to interpret. Figure 6.2 shows an example of how students might label the graph. Ask students to create an equation in slope-intercept form that represents the graph. For this graph, the equation is $y = 2x - 1$. In this case, students may also question how the graph representing increasing drone population begins at $(1, 1)$ and does not extend into the third or fourth quadrants. Students may need support to understand that the graph does not show a representation less than $(1, 1)$ because that point represents Day 1 when Cassie saw one drone and there has been a linear increase in the number of drones Cassie sees in the sky each day since.

On the other hand, Cassie provides the reader much numerical information about how the four previous waves impacted human population (pp. 44–45), and at one point she even exhorts the reader, "You do the math" (p. 45). We know from Cassie's account that the third wave caused the death of 97% of 4 million people in 12 weeks through a virus similar to the bird flu engineered by the aliens (p. 45). This specific instance does not fit this algebraic topic well though, because viral spread is well-known to follow nonlinear patterns. However, Cassie's account is less specific about how the first wave's technological impact brought about the death of half a million people, but it does tell us the time span was merely seconds. We are left to assume the failure of technology resulted in airplane crashes and other similar catastrophic events. The graph included in the example task could, in our judgment, address a 10-minute time period. Figure 6.3 shows an example of how students might label the graph. Students might label the x-axis in units of seconds based on Cassie's interpretation of the timing of the first wave, though. Students should be asked to create an equation in slope-intercept form that represents the graph provided. For this graph, the equation is $y = -0.05x + 7.5$. Students

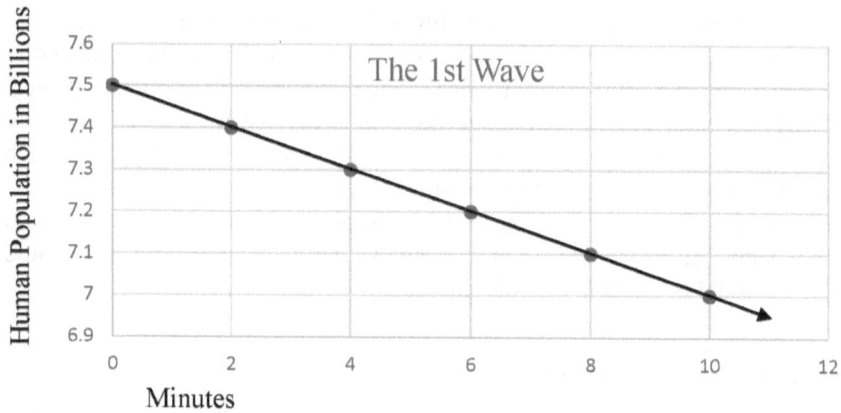

Figure 6.3 Labeling of First Wave Population Decline. *Source*: Created by authors

should also be encouraged to discuss why the line has an end point at (0, 7.5), not extending into the second quadrant. Students may be able to draw from their previous work with the graph of increasing drones over time to understand that the point (0, 7.5) represents the human population in billions on the first day at the beginning of the first wave. In this case, it is important for students to reason that there was a human population before the arrival of the aliens, but this graph is meant to be only a representation of what happened from the beginning to the end of the first wave.

Week 3: Linear Equations and Inequalities in Two Variables

Ben, who has named himself Zombie, learns from Dr. Pam in the training center that one in three humans left alive on Earth is not human at all, but rather a victim of an alien body takeover. He has no way of knowing which humans are legitimate and which are aliens but he is told that the other teens with whom he is training can be trusted. This week's reading offers opportunities to help students make connections between ratios and direct variation as it appears in equations in slope-intercept form when the y-intercept is at the origin. One suggested task for this week is as follows. Ask students to consider the following dilemma: Ben will travel with a group of his fellow teens from the training center to an outpost of survivors to bring food and water. Ben knows one of three people he meets may be an alien. Ask students to construct an equation or inequality in slope-intercept form that represents how many people Ben meets among the survivors who may be aliens. Then have students construct an equation or inequality in slope-intercept form that represents how many people Ben meets among the survivors who may be human. A student who chooses to construct equations should generate $y =$

$(1/3)x$ to represent aliens masquerading as human survivors and $y = (2/3)x$ to represent human survivors. However, students should recognize at this point that Ben cannot be sure if exactly one in three humans will be alien imposters in any situation, so he may need to think in terms of "at least" or "no more than." Some students may choose to represent aliens with an inequality such as $y \geq (1/3)x$. Check to see if they pair that inequality with $y < (2/3)x$ to represent humans. The proper pairing of inequalities should generate discussion on the meaning of the inequality symbols and slopes in this case. Students should also be asked to explain why the intercept is zero in this context.

Week 4: Systems of Equations

Like Cassie, Ben has also become aware that the small group of surviving humans is decreasing in numbers by the day. Ben also knows that each day more and more aliens emerge into awareness in human bodies. At some point, as humans kill each other and as more aliens take over human bodies, the population of humans and aliens will be the same. This is an opportunity for students to work with a system of equations in which one equation represents the decreasing human population and the other represents the increasing alien population. *The 5th Wave* does not provide much mathematical information beyond the initial population of 120 million surviving humans. However, this presents a dilemma similar to those proposed in three-act math tasks (Meyer, 2011). Three-act math tasks ask students to consider a task that can be thought about in an intuitive way, but is missing information or will need to be approached in an efficient manner (Hallman-Thrasher et al., 2018). In this case, students have access only to the number of humans. Encourage them to consider what they know. The population of aliens must start at zero and is increasing. The population of humans is decreasing. It is not known if the changing populations of humans and aliens can be modeled by a straight line, but for the purposes of this task, ask students to assume both populations are changing at a constant rate. This presents the opportunity for students to see that the slope in equations of lines represents constant rates of change. Ask students to draw two lines on a graph and then locate the point at which the human and alien populations will be the same. Students should generate qualitative graphs similar to that shown in figure 6.4.

Once students generate graphs, ask them to write the story of the graph. In writing the story, they should answer the following questions: *How many days, weeks, or years does it take for the human and alien population to be the same? What population is this? Which axis shows time, and in what units? Which axis shows population? Which line on the graph represents the increasing alien population and which line represents decreasing human population? What is the rate of change for the increasing alien population?*

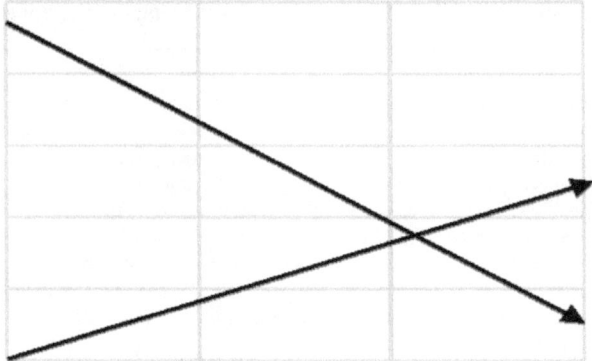

Figure 6.4 Week 3 Student Graph Example. *Source*: Created by authors

What is the rate of change for the decreasing human population? What were the starting populations for aliens and for humans? What equation of a line in slope-intercept form represents each line? Keep in mind that students' graphs, stories, and equations will be different. The point of commonality of all graphs and equations should be a y-intercept of 120 million for the human population and a y-intercept of 0 for the alien population.

Week 5: Rearrange Formulas to Highlight a Quantity of Interest

Ben is entrusted with information from Commander Vosch that explains why the soldiers are children and teens, and how they believe the aliens are trying to defeat the humans. Vosch tells Ben one major strategy is to cut off supplies from the training camp, causing the soldiers in training to starve. Present students with this formula: $C = -rT + I$. In this formula, T represents time in days, C represents cans of beans available to feed the soldiers at the end of a certain amount of days, r represents the rate at which soldiers are consuming cans of beans (in cans per day), and I represents the initial number of cans of beans in the camp. Ben's main concern with supplies is how long they will last. Have students develop a formula that will be more useful to Ben by solving it in terms of T, time in days. Students should arrive at this rearranged formula or its equivalent: $T = (I - C)/r$. After students work with rearranging the formula, facilitate a follow-up discussion about how Ben might use the formula. Table 6.2 presents some discussion questions along with potential answers.

Week 6: Function Concepts and Function Notation

Ringer, who is one of the soldiers under Ben's command, is the first to figure out that they are the fifth wave. She explains to Ben that Vosch and others

Table 6.2 Potential Discussion Questions to Support Using the Rearranged Formula

Discussion Questions	Potential Answers
Ben finds out there are 22,000 cans of beans in the camp and soldiers are consuming beans at the rate of 200 cans of beans a day. How long will it be until there are 1000 cans of beans in the camp?	Students should realize that they can use the rearranged formula to solve this problem. Ben wants to know time until there will be 100 cans of beans remaining in camp and the rearranged formula can give this information in days. The initial number of cans of beans (I) is known to be 22,000. Soldiers are consuming beans at a rate (r) of 200 cans per day. Ben wants to know how long it will be until there are 1000 cans of beans available (C) to feed the soldiers. The formula with substitutions is as follows: $T = (22{,}000 - 1000)/200 = 105$ days
How long will it be until there are no cans of beans in the camp?	Using the same rearranged formula, the initial number of cans of beans ($I = 22{,}000$) and the rate ($r = 200$) soldiers consume beans will not change. Since Ben wants to know how long it will be until no cans of beans remain, $C = 0$. The formula with substitutions is as follows: $T = (22{,}000 - 0)/200 = 110$ days
How might this impact Ben's perspective on Vosch's declaration that "We take the battle to them" (Yancy, p. 260)?	Students will need to think beyond the formulas to answer this question. Vosch is giving Ben a reason to believe remaining in camp is not an option. To remain longer than a few months means starvation. Leaving camp means fighting. Students may understand that Ben sees this as preferable to starving in camp.

like him are training more and more human soldiers to hunt and kill other humans, not aliens in human bodies as they have been told. In this part of the unit, students are exploring functions and there is a parallel between the literary plot and the mathematics being studied here. Functions are often thought of as machines that map inputs to outputs. The training camp where Ben, Ringer, and the other children and teens have found themselves can be thought of as a machine that converts children and teens into soldiers. In this way, the training camp serves a similar purpose to a mathematical function by converting inputs to outputs. Guiding students to see the parallels between the training camp and functions can help students make sense of the meaning of the mathematical concept of a function. It may also help students distinguish the difference in meaning between equations and functions. Assist students in building an understanding of equations as mathematical tools that are used to represent equality between expressions, while functions are used to represent relationships between sets of inputs and their corresponding outputs. Show students that because of this difference, functions use a different notation format than equations. Students could examine a function similar to this: $f(k) = 0.8k$, where k is the number of kids (children plus teens) entering the training camp and $f(k)$ is the number of soldiers produced from the kids. The students will notice that there are less soldiers produced from the function machine than kids that are input. Ask students what mathematical aspect of the function causes this to happen. Also, ask students what is happening at the training camp that supports the proposed function. Students may notice that it is the slope, which in this case has a value of 0.8, that mathematically changes less than the full number of children into soldiers. To answer the second question, students may connect the mathematical model to the deaths of Poundcake and Flintstone during the training exercise and realize that children do not always survive the exercises intended to convert them into soldiers. Would they propose a different function to better match the circumstances? If so, what would that function be and why?

Week 7: Interpreting, Comparing, and Graphing Functions

Cassie finally understands Evan's true nature and she has a very difficult conversation with him. One of the ideas they confront is what would humans do if they were the ones who could not survive on Earth and had to leave. This brings up a parallel dilemma with which students are likely to be very familiar, humanity's challenge in confronting climate change, a challenge that students will be quick to understand will not be solved by relocating humans to a new planet. Have students examine graphs that represent global warming trends over time. Have students review resources located on the Global Climate Change: Vital Signs of the Planet website developed by NASA (Shaftel, Jackson, Callery, & Bailey, 2021). For example, one graph found on the website, *Scientific Consensus: Earth's Climate Is Warming*, shows rapid warming trends in degrees Celsius from 1880 to 2020. The function $T(x) = 0.10x + 0.05$ has been used to approximate temperature data after 1960 where x represents decades since 1960 and $T(x)$ represents change in temperature in degrees Celsius since 1960. Have students use a graph to obtain average global temperature in 1960 and the function to find the expected average global temperature, if current trends continue, for 2030, 2040, and 2050 (Odenwald, n.d.). While the Global Warming Trends activity includes instructions for students to create a graph and predict the change in temperature from 1960 to 2000 and from 1960 to 2050, our activity proposes that students work with data on an existing graph and the provided function to predict future expected average global temperatures.

Week 8: Arithmetic and Geometric Sequences

When Cassie and Ben meet again, she has a secret she wants to tell Ben. Pose the following scenario to students based on the secret Cassie knows:

> After an hour Cassie learns that Ben already knows the secret but has not told anyone. Now they want to share the secret with others but they will have to be careful. What if Cassie and Ben each tell one person the secret after another hour, and then every hour a person who knows the secret tells someone new? How many people would know the secret after eight hours?

The spread of this secret follows a geometric sequence and students can model the sequence visually or through the use of the equation, $g_n = g_1(r^{n-1})$, where g_n is the value of the *n*th term in the sequence, g_1 is the first value in the sequence, and *r* is the common ratio or number used to multiply each successive term to find the next term. Students can model the sequence by creating a pattern like the one shown in figure 6.5 to discover that the number of people who know the secret doubles every hour.

From there it is relatively simple to determine that 32 people will know the secret after the 5th hour, 64 people after the 6th hour, 128 people after the 7th hour, and 256 people will know the secret after the 8th hour. Substituting into the equation $g_n = g_1(r^{n-1})$, to solve for g_8, where g_1 is equal to 2 and *r* is equal to 2 will yield the same result of 256 people.

Ask students what will happen if the secret continues to spread for the next eight hours in the same manner? Some students may incorrectly believe the number of people who know the secret after the next eight hours will only double while other students may be amazed at how many people will know the secret after another eight hours. Substituting values into the equation to solve for g_{16} this time students will find the following:

$$g_{16} = 2(2^{16-1}) = 2(2^{15}) = 65{,}536 \text{ people will know the secret after another eight hours.}$$

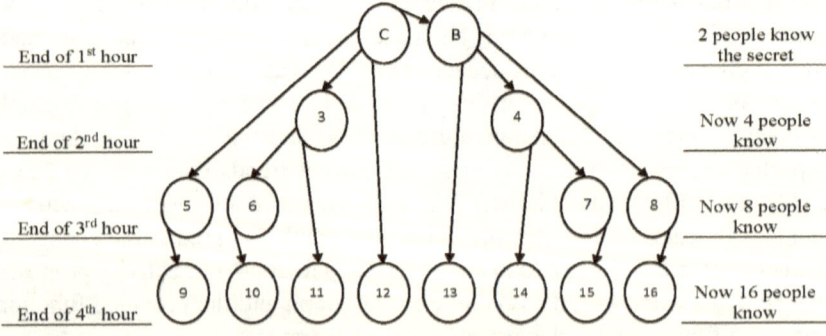

Figure 6.5 **Example Model of Geometric Sequence.** *Source:* Created by authors

AFTER READING *THE 5TH WAVE*

Expanding Prior Activities

Activities done after completing this cross-curricular unit are designed to give students the opportunity to synthesize their mathematics and literacy learning. During the reading of the text, students engaged with mathematical representations such as graphs and equations that connected with events in the story. After completing the reading and math activities, students might take up the question of literacy in mathematics by choosing one of the math activities and using that to represent the information in one of the nonfiction texts. For example, students might take the inequalities activity from Week 1 of the unit and the nonfiction text about water shortages on the Navajo Nation (Calma, 2020). Could they use inequalities to share the same information as the article? This may also lead to a discussion about variables; how many unknowns are there in a problem about water shortages? Students may recognize that we need to know the number of gallons of water an average family uses in a day when they do not have running water. They may need to look at the space needed to store water, or the economics of having to travel many miles to purchase water for use at home.

Adding Mathematical Equations to Section Titles

A second suggested post-reading activity involves the foreshadowing that Yancy does when he titles different sections of the book, such as "IX A Flower to the Rain" on page 311. Students might take the different section titles and add a mathematical equation, inequality, or graph to each with minimal labeling and then explain why they chose that mathematical representation to foreshadow the coming action.

BEYOND *THE 5TH WAVE*

Creating a Dystopia

After reading *The 5th Wave* and completing mathematics and literacy activities connected to the story, we suggest that students create their own dystopia using the concepts they have learned. We recommend that each student should include all of the elements of a dystopia that the class defined during the before-reading activity. They should also make connections to a minimum of three real-world topics and at least three mathematical concepts that were explored during the reading of *The 5th Wave* in their dystopias. Teachers might consider having students use other Disney or Pixar movies, such as *Star Wars* (Lucas, 1977) or *A Bug's Life* (Lasseter, 1998), as a starting point for this activity to add structure to the assignment and make an additional connection back to our before-reading activities. Student-created dystopias could be presented in many different ways, from something as simple as written responses, to a series of questions, to comic books, to video presentations.

Math in History

The main characters in *The 5th Wave* were forced to be resourceful and innovative in order to survive in their dystopian world. As an extension activity to combine both mathematics and literacy, teachers and students may want to research the development of various mathematical concepts during times of social upheaval. An example of this is Newton inventing calculus during a plague in the seventeenth century. Another example might be the development of number systems and arithmetic in ancient Mesopotamia when groups were joining together into larger cultural units that contributed taxes and such for the common good. This activity would be an excellent way to highlight the contributions of underrepresented groups in mathematics as well.

CONCLUSION

The events portrayed in *The 5th Wave* provide a meaningful context for mathematics teaching and learning. By combining the mathematical concepts and the novel, along with relevant nonfiction pieces, students have the opportunity to make real-world connections among the topics. What we have presented here is only a sample of the possible mathematics and literacy connections available with this novel. We highly recommend that mathematics and ELA teachers consider using this book as a cross-curricular event.

REFERENCES

Boissoneault, L. (2017). *The true story of brainwashing and how it shaped America*. The Smithsonian. https://www.smithsonianmag.com/history/true-story-brainwashing-and-how-it-shaped-america-180963400/.

Brozo, W. G., & Crain, S. (2018). Writing in math: A disciplinary literacy approach. *Clearing House, 91*(1), 7–13.

Burke, J., & Hatcher-Moore, P. (2017, July 24). If you are old enough to carry a gun, you are old enough to be a soldier. https://www.theguardian.com/global-development/2017/jul/24/south-sudan-child-soldiers.

Calma, J. (2020). *The Navajo Nation faced water shortages for generations—and then the pandemic hit*. The Verge. https://www.theverge.com/2020/7/6/21311211/navajo-nation-covid-19-running-water-access.

Collins, S. (2008). *The hunger games*. Scholastic Press.
Darabont, F., & Kang, A. (Producers). (2010–2021). *The walking dead* [Television series]. AMC.
Davis, E., & Griggs, B. (2019). *The world's population is nearing 8 billion. That's not great news*. CNN. https://www.cnn.com/2019/07/11/world/world-population-day-trnd/index.html.
Donahue, D. (2003). Reading across the great divide: English and math teachers apprentice one another as readers and disciplinary insiders. *Journal of Adolescent & Adult Literacy, 47*(1), 24–37.
Draper, R. J. (2002). School mathematics reform, constructivism, and literacy: A case for literacy instruction in the reform-oriented math classroom. *Journal of Adolescent & Adult Literacy, 45*(6), 520–529.
Hallman-Thrasher, A., Koestler, C., Dani, D., Kolbe, A., & Lyday, K. (2018). Graphing stories for a three-act task. *Mathematics Teaching in the Middle School, 24*(2), 90–96.
Lasseter, J. (Director). (1998). *A bug's life* [Film]. Walt Disney Pictures and Pixar Animation Studios.
Lowry, L. (1993). *The giver*. Houghton Mifflin Harcourt Publishing Company.
Lucas, G. (Director). (1977). *Star wars episode IV: A new hope* [Film]. Twentieth Century Fox.
Meyer, D. (2011). *The three acts of a mathematical story*. dy/dan (blog). http://blog.mrmeyer.com/2011/the-three-acts-of-a-mathematical-story/.
Miller, B. (Producer). (2017–2021). *The handmaid's tale* [television series]. Hulu.
Muhammad, G. (2020). *Cultivating genius: An equity framework for culturally and historically responsive literacy*. Scholastic.
National Council of Teachers of Mathematics. (2014). *Principles to actions: Ensuring mathematical success for all*. https://www.nctm.org/PtA/.
Odenwald, S. (n.d.). *Space math @ NASA*. National Aeronautics and Space Administration: Goddard Space Flight Center. https://spacemath.gsfc.nasa.gov/SpaceMath.html.
Orwell, G. (1949). *1984*. Harcourt, Inc.
Ross, G., & Lawrence, F. (Directors). (2012, 2013, 2014, 2015). *The hunger games* [Film series]. Color Force and Lionsgate Films.
Shaftel, H., Jackson, R., Callery, S., & Bailey, D. (2021). *Scientific consensus: Earth's climate is warming*. NASA: Global Climate Change: Vital Signs of the Planet. https://climate.nasa.gov/scientific-consensus/.
Spielberg, S. (Director). (2005). *The war of the worlds* [Film]. Paramount Pictures and DreamWorks Pictures.
Stanton, A. (Director). (2008). *WALL-E* [Film]. Walt Disney Pictures and Pixar Animation Studios.
Yancey, R. (2013). *The 5th wave*. G.P. Putnam's Sons.

Chapter 7

"Absolute Zeros Solve for Why" Using the Pythagorean Theorem in *Island of the Unknowns*

Brian Rothbaum and Julie Grasfield Weil

Island of the Unknowns (Carey, 2009) is an enigmatic title, as the reader at first does not know whether the unknowns are things or people. Di Smith and Tom Jones, the unlikely heroes of the novel, could not have more generic last names. But, as the mathematics and the mystery both unfold, the reader discovers the depth and breadth of these characters and their inner strength and resourcefulness, thanks to a teacher who believes in their intelligence and persistence. Just as students solve the mathematical mystery of the unknown posed in the novel, so too will they unravel the mystery of characterization, as the adolescent protagonists are developed through their words, thoughts, actions, and observations.

As George Pólya, the posthumous character in the novel, but very real mathematician and author of *How to Solve It: A New Aspect of Mathematical Method* (1945), observed, "Solving problems is a practical skill like, let us say, swimming. We acquire any practical skill by imitation and practice" (p. 4). *Island of the Unknowns* introduces several major mathematical concepts that are conducive to more in-depth study. Deciphering clues interspersed throughout this young adult novel provides students wonderful opportunities for classroom investigation of three major topics in mathematics: the Cartesian coordinate plane, Pythagorean Theorem, and probability.

Through this mystery novel, teachers can present the principles of Euclid, Pythagoras, and Descartes, so that these mathematical giants can be brought to life in the classroom and their works explored. Using various pedagogical methods demonstrated in the novel, students will be transported from mundane math problems to real-world applications.

A NOTE TO TEACHERS

Polya (1945) gave four logical steps in order to solve a problem. These can be applied to both mathematical problems and character development:

> First, Understanding the problem: *What is the unknown?* (emphasis in original) *What are the data? What is the condition?* ... Draw a figure.... Second, Devising a plan: Find the connection between the data and the unknown. ... Third, Carrying out the plan: ... *Check each step.* ... Fourth, Looking back: Examine the solution obtained. Can you *check the result?* ... Can you derive the result differently? (1945, pp. xvi–xvii)

The activities suggested in this chapter will leverage these steps in guiding students through mathematical problem-solving.

ISLAND OF THE UNKNOWNS BY BENEDICT CAREY

Di Smith and Tom Jones, key characters in the novel and relative ciphers in their community, are determined to find their beloved tutor Mrs. Clarke, after she is the third member of their tight-knit community to disappear. They live in a mobile home park called Folsom Adjacent, which is literally and figuratively next to nothing, and was named after the nearby nuclear power plant. The pair discovers that Mrs. Clarke left them cryptic clues in her mobile home—stirring straws and a salt shaker to represent a math equation—that only they can decipher. Enlisting the help of some other teens, each with his own skill set, Di and Tom set out on a quest to find Mrs. Clarke and unlock the mystery.

Guided by the clues, this bunch of unknowns solves for an unknown using the Pythagorean Theorem, and other math strategies. As the story unfurls, the teens discover their individual talents, that when used in concert, unravel the mystery. They devise schemes and create maps to help them find their way to a solution. They gain confidence, so that eventually these zeros become heroes.

BEFORE READING *ISLAND OF THE UNKNOWNS*

Mrs. Malba Clarke, the teens' favorite teacher and the fictitious embodiment of the real George Polya, advised her students, "Simplify. Take the simplest version of the problem and solve that. Then try the harder ones" (p. 155). George Polya's Step 2 directs one to devise a plan, guiding one to discover a connection between the data and the unknown. Help students preview the novel with a challenge in the form of a great-escape box full of items and clues to study and interpret to familiarize students with a sampling of items addressed in the novel and to stimulate wonder. It represents a simplified version of the larger problem readers will encounter with the mystery of the novel, and it provides tools to help them find the solution.

THE GREAT ESCAPE

Engage students by telling them that the class is going to embark on an escape-the-room adventure. Pique their curiosity and prepare them to scour for clues. Explain that they'll need to pay close attention to detail regarding character descriptions and mathematical expressions. Divide students into teams of four or five. Supply each team a box with a facsimile of a three-digit combination lock and a cryptic note from "Mr. E" alluding to the contents:

> In this letter, I have included a number of things that will help you on your journey. Just don't have tunnel vision. If you can figure out the clues, you'll find your way to what's right. It should be easy as pie. Try angel's food or pineapple upside-down cake. Patience builds character.—Mr. E.

Before opening the box, ask students to decipher the signatory. Note how each team goes about puzzling over possible combinations. Once the team correctly guesses "mystery," depicted as "3JW" (the three symbols upside-down), the teacher should immediately go over and "open" their box. If there are teams that are unable to decipher the signatory, after a few minutes, walk around and "open" all the locks with your "key."

Once all teams' boxes have been opened, ask students to take out and observe the contents of the box: a navigational compass; a mathematical compass; a ruler; a copy of the tide table from the novel (p. 123); a notepad and pencil; laminated paper shapes—triangle, circle, square, rectangle; a copy of the novel; a laminated paper with an "x" on it, one with a handwritten rounded "E," "W," "M," or even a "3," a handwritten "h" or "4," a "Z" or "N," an "I" or "H," and a "p" or "d" (to show them that sometimes they have to look at things in different ways); and a laminated copy of the map from the book (p. ix). Encourage students to hone their detective skills by unpacking all of the contents of the great-escape box and laying them out on a table in perceived categories. Silently observe if any groups analyze the wording of the note. See if they turn the letter/number cards around to view them in different ways.

This activity can be revisited after reading the novel and see how many noticed all of the clue words embedded in the note: *letter, number, tunnels, clues, right, pi, Angels (baseball cap) upside-down, characters upside-down,* and *mystery*.

CARTESIAN COORDINATE PLANE, PYTHAGOREAN THEOREM, AND PROBABILITY

Prior to delving into the book, it is important that students preview the three major mathematical concepts—the Cartesian coordinate plane, Pythagorean Theorem, and Probability that will be presented within the text. Lessons should be structured to offer students opportunities to wrestle with problem-solving tasks without direction, encouraging them to develop their own approach to problem-solving, mirroring the characters' actions in the book.

EASY AS 1, 2, 3—ER, 3, 4, 5

The author includes explanatory examples of Pythagorean triangles in images on pages 26 and 28. He even illustrates how the numerical square on each side is a literal square. Give students practice with the formula. Write $a^2 + b^2 = c^2$ (the Pythagorean Theorem) on the board. Place a jar of Popsicle sticks on your desk with correct and incorrect combinations for Pythagorean triples written on them for students to take. Possible combos: (6–8–10), (9–12–15), (5–12–13), (8–15–17). Demonstrate with (3–4–5) that works and (3–4–6) that doesn't. Have students choose a Popsicle stick and plug the numbers into the formula to see if each is a viable triple. Then, take another Popsicle stick. The first team to find five correct Pythagorean triples wins.

PLANE AND SIMPLE

The coordinate plane is a familiar mathematical concept. Reinforce students' understanding of the physical layout of a Cartesian plane and how to plot coordinate pairs on it. Create a Cartesian coordinate plane for students to navigate. Tape off the classroom with x and y axes. One could even follow the lines of floor tiles. Clearly mark the origin at the intersection of the x and y axes in the center of the room. Hand out cards with coordinate pairs. Remind them that $(0, 0)$ is the origin, where the x and y axes intersect. Have students start at the origin and walk to their coordinates. String yarn between points to indicate lines. Be sure to create an image from all successfully mapped lines of the designated coordinate pairs.

ROLL OF THE DICE

Let students investigate the probabilities of numerous dice rolls. Divide the class into groups of four. Provide each group with two dice and ask them to try to figure out the probability of rolling snake eyes (sum of 2), a 7, a number greater than 10, and a 12. Ask students, "How many outcomes are there, and why?" Have them note their results on a "Sum of 2 Dice" outcomes chart, a 6 × 6 grid. Note the equivalent probabilities of certain outcomes, such as 2 and 12 and 3 and 11. The probability of rolling a sum of 2 is 1 in 36, which is the same as the probability of rolling a sum of 12. Point out visual patterns in probability and review the fundamental counting principle.

MAGICAL MYSTERY TOUR

Before acquainting students with the characters, first lay the foundation by steering students through a thorough examination of various aspects of the novel, from cover to cover.

Guide students on a book walk in order to familiarize themselves with helpful illustrations and writing. Prompt students to study the cover, the chapter titles, the map on page x, Mrs. Clarke's journal entry on page xi, the back cover, and then flip through the pages to view the illustrations. Encourage students to discuss observations in small groups and see what insights they glean. Ask each group to jot down questions and comments on these diverse features of the novel, and then invite them to share three with the whole class. Have them tuck these away for now, but note any overlap so these can be addressed further when students come across them in the novel. Examples may include the following: a star, eights, and circles on the cover; notable chapter titles, *The Outhouses*, *Thunder Underground*, and *The Amulet*; and memorable illustrations, the squares (p. 22), the neighborhood (p. 37), and the outhouse (p. 66).

WHILE READING *ISLAND OF THE UNKNOWNS*

Just as Mrs. Clarke incorporated her students' math problems into the fictious tales she wove, so too will students engage in learning as they read *Island of the Unknowns*. By introducing the Pythagorean Theorem, the

Cartesian coordinate plane, and probability in the context of the novel, students are instantly hooked before they realize that they're learning mathematical concepts. They willingly join Di and Tom on their quest to find out what malice has befallen their favorite tutor and the local nuclear plant.

WHAT'S YOUR ANGLE?

Peppered throughout the story are words of advice from Mrs. Clarke about working through problems:

> When you can't decide whether to act or wait, it is usually better to act. Acting sets the mind in motion, and you can always change direction if you're wrong. And searching for a solution is the best reminder that there is one. (p. 9)

And, "Simplify. Take the simplest version of the problem and solve that. Then try the harder ones" (p. 155). Have students set up sleuthing notebooks and direct them to note Mrs. Clarke's words of wisdom as they read the novel.

Explain that Mrs. Clarke is a fictitious version of George Polya. Share his four-step plan for problem-solving. Direct students to apply these to real-world mathematics examples. Fermi questions lend themselves to this exercise: *How many jelly beans are in the jar? How many pizzas does our class eat in a year? How old are you if you're a million seconds old?* Solve one together as a class. Then, assign others to groups. Reconvene and discuss methods.

CHARACTER COUNTS

Characterization is a literary device that reveals characters' thoughts and actions in a step-by-step manner. Direct characterization is when the author clearly states a character's physical appearance and personality. Indirect characterization is when the author depicts what the character does, says, and thinks while also showing how other characters react to him. For example, Di is analytical, proposes theories, and often quotes the wisdom of Mrs. Clarke. Tom seems shy and far less willing to speak up, but he's always pondering the clues. The author helps to flesh out these characters by portraying how others interact with them. Through exploring the ways by which an author develops character, students can calculate the value of certain qualities and behaviors.

Advise students to note down every observation they discover about Di and Tom, as the novel progresses. Have students set up characterization charts with the following categories in their sleuthing notebooks: physical descriptions, thoughts and feelings, actions, dialogue, and other characters' comments about them. Go from individual notes to sharing in pairs to sharing in small groups. Then, reconvene as a class to share details.

Carey starts with physical descriptions of his two protagonists:

> Di's real name was Diaphanta . . . Smith. . . . Di was all right. . . . She had long, orange hair and this habit, kind of like a tic, where she kept twirling her right wrist, like she was working out a cramp or something. (p. 4)

Her best friend Tom Smith's moniker simplifies a lengthy name.

> His full name was Tamir Abu Something Something al-Khwarizmi. . . . They just called him Tom Jones. . . . He was tiny for an eleven-year-old, bony as a little bird, and you never saw his eyes. He wore this Angels baseball hat all the time, everywhere, pulled down low. (p. 5)

From these excerpts and the text of Chapters 1 to 4, readers can infer that Di is nervous and observant, while Tom is shy and contemplative. Di likes to use manipulatives to work through math problems, while Tom would rather visualize and ruminate. They are loyal friends and quite fond of their favorite math tutor Mrs. Clarke. We'll discover later that Di is overweight. For now, we know she's a redhead and that Tom is a small brunette who hides behind an Angels baseball cap.

GRASPING AT STRAWS

As students read Chapters 3, 4, and 5 in the novel, they will recognize the universal constant of the Pythagorean Theorem by constructing a 3–4–5 triangle out of straws. Distribute drinking straws cut into uniform lengths and building bricks, and let students demonstrate the Pythagorean Theorem, illustrated in the book on pages 22 to 28. Provide students an opportunity to work in groups with manipulatives to simulate the straw figure on page 26. Point out how the 2^2 is a "square" and can be depicted as a square shape with an area of four units. Ask students to configure the squares and then create with their manipulatives the Pythagorean triangle in the diagram on page 28. Build on the 3–4–5 Popsicle-sticks lesson in the Before-Reading section, by challenging them to graduate from 3–4–5 to 5–12–13 triangles.

COORDINATING TRAFFIC SIGNALS

This activity aligns with Chapters 7 to 10. Di and Tom discover the starting point of their descent into the tunnels, from Mrs. Clarke's clue to them indicating the origin coordinates (0, 0). See the illustration on page 66. The idea of mapping is introduced in Chapter 10 and then evolves into the coordinate plane. The concept of the coordinate plane can be examined in the classroom by having the students play a Battleship-like game with cars on the plane. Have students reread Chapter 7 and get a sense of the coordinate pairs (0, 0) and (−12, 8). Pass out paper playing "boards" with a grid of New York City streets, with the central point (0, 0), and five vehicles drawn in different locations at whole-number coordinates. Let pairs of students face each other, call out coordinates, and see who's first to locate all of their opponents' vehicles: motorcycle, sedan, SUV, school bus, and 18-wheeler truck.

EUREKA!

Just as one sets up an equation by noting what one already knows and identifying the unknowns, so too does one identify the literary elements of a novel and the plot. Di and Tom gather clues and collect resourceful people. Discuss with the class one or two early on, such as Ham in Chapter 10, which can be added to their sleuthing notebooks. Tell students to cite passages where they make predictions, seek knowledge, and learn useful information. Also, ask them to show evidence for how and when Thea, Ham, and Oki change and become integral members of the team.

ZEROS TO HEROES

Ask students to fill out as many details on their characterization chart as possible as they read Chapters 16 to 18: physical descriptions, speech, thoughts, actions, and others' observations. Recommend that they note details about Di and Tom's friends Ham, Thea, and Oki described here. In class, allow students time in small groups to discuss their findings and add to their own charts. Post around the classroom kraft paper with the various headings from students' characterization charts noted earlier, hand each group a different color marker, and ask them to contribute one entry to each poster. Then, lead a class discussion on what clues were found, and how this scene constitutes a pivotal moment in the novel. Analyze how the group collaborated to decipher mathematical clues, calculate distances, and navigate through the underground tunnels.

CRACKER JACK

Every good caper has a code to crack. Oki leads the gang to figure out the probability of the password code. Read Chapters 19 to 20 in class, stopping periodically to have students analyze the figures and hazard guesses as to the hidden numbers and the meaning of Big Sip's clue. Then, read on for the answers.

Read Chapter 21 aloud as a class, and work through setting up the probability problem to figure out the password together. Have students compute the total possible number of permutations, as explained in Oki's step-by-step process. Highlight Oki's explanation starting on pages 214 to 215:

> Assume that 3–4–5 came as a package, in that order. If so, that trio could be in only one of three places. If it was at the beginning, that left two open spaces at the end. Each of those two spaces could have any number from 0 to 9 in it. (p. x)

Follow through the entire process of finding the possible combinations that can be derived by using this algorithm.

Explore Polya's advice with students to start with a simpler, familiar problem and let them practice this skill. Encourage students to explore the fundamental counting principle in different teams, determining outcomes for sandwich breads, cheeses, and meat choices; pizza crusts, cheeses, and toppings; clothing jackets, shirts, and pants; and car models, colors, and upholstery. Model these probabilities with a tree diagram. How many combinations can you come up with for each scenario? As a class, compare findings. Discover that the probabilities are all parallel.

AFTER READING *ISLAND OF THE UNKNOWNS*

Once students have explored the concepts of the Pythagorean Theorem, the coordinate plane, and probability, and studied characterization through the lessons while reading the novel, offer students a variety of culminating activities to exhibit mastery.

Lights, Cameras, Action!

Have students create a movie poster for this novel. Let them demonstrate their acquired knowledge by having them choose actors for major and minor roles, describe how the protagonists evolve throughout the story, depict a crucial scene through artwork, cite a quintessential quote and identify who said it and explain its significance, write a tagline, pen a critic's review, choose a theme song or one to play as backdrop for a particular scene—and explain why they chose it.

Chart Your Course

In order to reinforce students' knowledge of the coordinate plane, engage them in exercises to create recognizable images from sets of coordinate points. Give students numerous points to plot along the x and y axes. Once they have mastered this skill, assess student comprehension by asking each student to create a complex set of mystery plot points that depict an image from the text when plotted correctly. Have students exchange plot points and connect one another's dot-to-dots. Bonus: note a plot point, or turning point in character development, in the novel.

Write Angles

Offer students a fun and creative way to demonstrate their recall of the events of the novel by allowing them to experiment with a wide array of writing styles. Assign each group of students to write different features of a newspaper: a column about Mrs. Clarke; an article on the police investigation into the disappearance of Mr. Romo, Mrs. Quartez, and Mrs. Clarke; an interview with Di and Tom; a column on teen heroism; a letter to the editor about living close to a nuclear power plant; a review of Tom's dad's kebab restaurant; a classified ad for homes in the Folsom Adjacent trailer park; and an ad for Folsom Power Plant employment.

The Aftermath

Readers often wonder what happens to the characters after the story ends. Invite students to write an epilogue tying up all of the loose ends of the story, with the characters and the town. Ask them to delete, change, or add a scene—and explain why.

Board Meeting

Challenge students in groups to devise a simple board game—based on Chutes and Ladders, Candyland, or even a checkerboard—that incorporates all three math concepts. It should be drawn on a Cartesian coordinate grid (10 × 10 squares), include a Pythagorean triangle, and deal with probability. It can depict the island or another locale. Markers can be the main characters, and the drawing cards can offer them challenges. Task students with creating 20 drawing cards that incorporate elements of the story, such as character traits of the main and ancillary characters, and details of places that have character: Mount Trashmore, the Folsom Adjacent trailer park, and Mrs. Polya's store. Another pile of 20 cards should involve math calculations regarding coordinates, right triangles, and probability. Students roll the dice to advance spaces on the board and to solve some probability questions, such as rolling a certain number with the dice, advancing to a specific square, and drawing a particular card. Allow some class time for other students to play these games with their creators.

BEYOND *ISLAND OF THE UNKNOWNS*

Challenge students to expand on their knowledge of these mathematical concepts beyond textbook problems in order to gain deeper understanding, starting with researching the real mathematicians alluded to in the novel and continuing with math exploration of advanced concepts. Each of these mathematical topics can be expanded upon endlessly. The Pythagorean Theorem is embedded in so many areas of math; it will become almost ubiquitous in its usage. The coordinate plane and graphing will be utilized by students through high school and into college. Probability is integral in most secondary and postsecondary math and science curricula. Likewise, encourage them to apply their newfound skills of analyzing characterization to other works of literature.

What's in a Name?

Arouse students' curiosity by inviting them to guess and then research the origins of characters' and place names: Examples: Di Smith's real name is Diaphanta, a nod to Diophantus of Alexandria, an Egyptian mathematician who wrote a series of books about solving algebraic equations; Tom Jones is the nickname given to Tamir Abu al-Khwarizmi, named for Muhammad ibn Musa al-Khwarizmi, the Muslim mathematician who wrote two influential books, one on algebra and one on calculation and doing arithmetic with the Hindu-Arabic numerals that he introduced to European mathematicians; and George Polya, Mrs. Polya's fictitious son who suffered an untimely demise, was actually a Hungarian mathematician who wrote a book of problem-solving techniques. Allow students to work in small groups and research one of the above names and or other famous mathematicians such as Pythagoras, René Descartes, or Blaise Pascal, and present a short oral report to the entire class.

Time Is Money

Teachers can expand upon the coordinate plane traffic-jam exercise, depending on student level and prior knowledge, to further explore René Descartes and his contribution to the concept by applying it to kids' first jobs. Linear equations and slope-and-intercept applications can all be explored from this jumping-off point. Linear expressions can be represented with application in the real world. Guide students to set up a chart with the y axis (pay—in increments of $15) on the left side and the x axis (hours) on the bottom. Have students chart the slope of their increased pay based on more hours worked.

Pythagoras's theorem is central to physics and mathematics, but scholars are unsure of its historical origins.

> We highlight a purely pictorial, gestalt-like proof that may have originated during the Zhou Dynasty. Generalizations of the Pythagorean theorem to three, four, and more dimensions undergird fundamental laws including the energy-momentum relation of particle physics and the field equations of general relativity, and may hint at future unified theories. (Overduin and Henry, 2019)

Fermenting Fermat

Fermat introduced the idea in a margin of Diophantus's book *Arithmetica* in 1637 and claimed to have a proof that demonstrated that no three positive integers a, b, and c satisfy the equation $a^n + b^n = c^n$ for any integer value of n greater than two. Since the text was discovered 30 years after his death and no proof was found, the question eluded mathematicians for centuries. The problem remained unsolved until Andrew Wiles published a formal proof in 1995 supporting Fermat's claim. Guide students to explore the mystery of Fermat's last theorem and allow them to experiment with the equation. After adequate exploration, tell them that the brilliant math minds behind many of the allusions on *The Simpsons* played a hoax on perceptive viewers. Show students the episode entitled, "The Wizard of Evergreen Terrace" (season 10, episode 2), where Homer seems to have written Fermat's last theorem on a blackboard, and have them apply their knowledge of Pythagorean triples to see if it's a possible solution. Point out afterward that it was a trick that used a rounding error that was due to the number of digits a calculator could display that allowed for this. Invite students to write a report on Andrew Wiles and his work.

Probabilities

The opportunities for extension activities for probabilities are endless. Strong lessons that draw from the book could relate to phone numbers, passwords, or license plates. Invite students to evaluate questions concerning how many options there are for combinations given certain limitations (must include particular digits/letters/symbols in passwords; phone number area codes cannot begin with 0 or 1; license plates can have 10 numbers and 26 letters in every place, etc.). Groups can be assigned the following tasks, and then they can report out their answers:

- The seven-digit numbers in a given phone number (within an area code, but not inclusive of the area code) that have the form ABC-XXXX, where X, B, and C can be any digit zero to nine and A is restricted to two to nine. There are two other restrictions: (1) B and C cannot both equal 1 since these values are designated for other purposes such as 911 (emergency) and 411 (information), and (2) 555-0100 through 555-0199 are reserved for fictional uses such as in television shows or movies. According to these conditions, how many 7-digit numbers are possible in a single area code? ($8 \times 10^6 - [8 \times 1 \times 1 \times 10^4] - 100 = 7,919,900$).
- New Jersey system of license plates has three letters followed by two numerical digits and one letter (e.g., AAA 11AA). How could you determine how many license plate numbers New Jersey has under this system? ($26 \times 26 \times 26 \times 10 \times 10 \times 26 = 45,697,600$).

The Power of Two

In small groups, challenge students to craft an outline of the sequel to this story. They should choose a mathematical concept to address. Or, include higher-level problems involving the current concepts. Perhaps incorporate characters with mathematically inspired puns for names such as Euclid, Ian; Pythagor, Ian; Bifour, Kate; Al Gore Rhythm; and George (Geo) Metrick. Design a chart to note the elements of characterization. Provide details for a few of these. Suggest a starting point of Mrs. Clarke's cryptic M $4 - 3 = 7$ message sign-off or let students devise their own.

CONCLUSION

Kathleen Turner's character Peggy famously said in the eponymous 1986 movie *Peggy Sue Got Married*, when she was magically transported 25 years back to high school algebra class, "Well, uh, Mr. Snelgrove, I happen to know that in the future I will not have the slightest use for algebra. And, I speak from experience" (Coppola, 1986, 01:43). Let's disprove that for our students and let them know that secondary mathematics classes help students develop critical-thinking skills so that they can successfully solve myriad problems throughout their lives. *Island of the Unknowns* does not provide rote math problems in textbook style, but instead offers adolescents real-world situations in which they can analyze a problem, devise a way to solve it, and apply the concepts of the Pythagorean Theorem, coordinate plane, and probability. Through the reading of this novel, desire for agency can be realized as students are encouraged to take the reins and plot their own course, literally and figuratively.

REFERENCES

Carey, B. (2009). *Island of the unknowns*. Amulet Books.
Coppola, F. F. (Director). (1986). *Peggy sue got married* [Film]. TriStar Pictures, Rastar Pictures, Zoetrope Studios, and Delphi V Productions.
Overduin, J., & Henry, R. C. (2020). *Physics and the Pythagorean theorem*. https://arxiv.org/pdf/2005.10671.pdf.
Polya, G. (1945). *How to solve it: A new aspect of mathematical method*. Princeton University Press.

Chapter 8

Exploring Math, Culture and Stories through *Math Girls*

Shelly Shaffer and Carlos Castillo-Garsow

Most young adult (YA) novels incorporate character hobbies, such as reading, art, music, or sports, but mathematics has seldom been explored as something young people do in their free time. This chapter shares approaches for using the YA novel *Math Girls* (2011) by Hiroshi Yuki in mathematics classes such as trigonometry, precalculus, or calculus. Because this novel integrates prose with mathematical equations, readers learn to code-switch between the mathematical and everyday register of English (Zazkis, 2000). Through this text, teachers can present advanced mathematical concepts as readers explore attitudes toward learning and doing mathematics by following several characters in the book as they complete sequences, rotations, geometric progressions, develop functions, and explore many other challenging mathematical concepts.

This book is a translated text. As such, special attention will be given to the impact that Japanese culture and language has on the story and text itself. Translators face obstacles when a text is rooted in Japanese culture and there are important lexical and syntactical differences between Japanese and English language (Hobbs, 2004). In this chapter, we offer approaches for students that follow the character arcs and examine narrative elements through a critical lens. Additionally, because the book was translated, some of the original intent was muddled in translation. Students can think critically about how the meaning of the story is impacted by language and culture.

CONNECTING MATHEMATICS AND LITERACY

The primary mathematics content of *Math Girls* is the study of sequences and series. As the characters work through challenging problems, they use the tools of precalculus to explore a variety of high school topics, including recursive and explicit formulas, the complex numbers and the fundamental theorem of algebra, polynomial generating functions, a proof of the binomial theorem, and the sine function. The author touches on some early calculus concepts, including continuity, limit, derivative, and Taylor series.

Yuki's characters struggle through their mathematical problems, and in doing so, exemplify many processes and proficiencies important in mathematics education. The characters also engage in mathematical practices that are key to understanding mathematics at all grade levels. They make sense of problems and persevere in solving them, use appropriate tools strategically when teaching each other, attend to precision in their discussion of those problems, construct viable arguments and critique each other's reasoning, and look for and make use of structure and repeated reasoning as their mathematical journey develops.

In addition to the opportunities for students to practice mathematics present in the text, students can also practice and master several literacy skills. Activities shared in this chapter ask students to analyze complex characters and particular points of view with particular attention given to the cultural experience reflected in the novel. Overall, students will analyze concepts presented in the text through two different mediums: mathematical equations and prose.

MATH GIRLS BY HIROSHI YUKI

Math Girls (Yuki, 2011) has been translated into English from Japanese. This book is the first in a series of novels focused on three high school characters whose hobbies are learning and exploring mathematics. Through the voice of an unnamed male narrator, *Math Girls* explores what it means to be "good" at math, while readers also learn about Japanese culture, experience the story of budding romance, and explore mathematics itself. The characters in the book figure out mathematical equations and concepts as a hobby. The narrator explains early in the novel that he views his hobby as unusual and does not fit in well with peers, but he finds a community of characters who share his passion for mathematics. The characters' interests in mathematics present a unique part of the book's plot. The characters primarily interact and develop their relationships through their collaborations on math problems. Each of the three main characters works through different levels of problems, which allow readers to engage the story at a mathematical level that is comfortable for them.

BEFORE READING *MATH GIRLS*

Activating prior knowledge and making connections is a key before-reading activity. In this case, connections to the text can be made through several individual, small-group, and whole-group activities that connect the text to mathematical concepts and Japanese culture.

Mathematical Prerequisites

Many of the chapters in *Math Girls* involve sequences and sequence notation. It is helpful if students have some experience with recursive notation for sequences before working some of the suggested activities. Before reading, students should be familiar with the idea that an arithmetic sequence generated by repeated addition of a common difference d can be represented recursively as $A_{n+1} = A_n + d$, which can be solved for the closed form $A_n = A_0 + dn$, and that a geometric sequence generated by repeated multiplication of common ratio r can be represented recursively as $G_{n+1} = rG_n$, which can be solved for the closed form $G_n = G_0 r^n$. Some experience factoring polynomials, even if just quadratics, is also necessary.

Additionally, several chapters of the book, and follow-up activities, involve combinatorial notation consisting of simple factorials, permutations, and combinations. Alternatively, the teacher may wish to delay teaching these topics until they are needed in Chapter 7, just as the characters in the book are often introduced to ideas and techniques only when a problem requires them.

Quick-Write

Quick-writes are an effective strategy to get ideas down on paper quickly. Students write for two to three minutes nonstop, concentrating on putting down all of their ideas, without censoring what they write or worrying about format or grammar. With a quick-write, students are able to explore ideas individually in response to a prompt. Quick-writes can be done prior to reading a text and then revisited after reading so that students can revise their ideas based on what they learned and experienced in the text.

For *Math Girls*, students should begin thinking about their feelings about mathematics prior to reading the text. Students can respond to questions such as the following: *What does it mean to be good at math? How do people know that they are good at math? What are the characteristics of people that are good at math? What might cause somebody to struggle at math?* After students write their responses, they can turn to a partner and read what they wrote. This will provide students time to share their opinions about mathematics with at least one other person in the class and compare. Have students save their quick-write to revisit after reading the novel.

Who Is the Narrator?

To build background knowledge and interest in the text, teachers could address the point of view and the narrator of the story. There are a few reasons the point of view matters in *Math Girls*: (1) this is a translated story where, in the original language, the gender, and point of view of the narrator is obvious; (2) this story takes on the Japanese characteristic of storytelling where the narrator is unnamed throughout the text; and (3) in the English version, readers do not know the gender of the main character until Chapter 4. The following section presents activities that lead students through an exploration of the narrator and point of view.

Prologue

Ask readers to explore the first two pages of the text from the "Prologue" (pp. 1–2). Students could read aloud this excerpt in small groups, with one person in each group taking on the role of the narrator. Because the gender of the narrator is not revealed in the text until Chapter 4, teachers might consider alternating among the various gender identities represented in each group as they read the excerpt aloud and should encourage students to read with expression. For example, when reading, "But math was too hefty a weapon for me in those days. I had only just got my hand around its hilt, and I wielded it clumsily—like I handled life, and my feelings for Miruka and Tetra" (p. 2), readers will be able to infer the gender of the narrator based on textual hints, but the first-person narrator remains a mystery. By reading the prologue, students will begin to infer characteristics of the narrator.

The teacher can have students use a graphic organizer (see figure 8.1) to guide students' predictions about the main character after reading the excerpt. Using graphic organizers has been proven as an effective reading strategy that offers students opportunities to engage with and think about complex texts (Marzano, Pickering, & Pollock, 2001). Tracking students' answers to the questions will permit students to see what other groups have inferred and compare their own predictions to the predictions of others. This will also encourage the class to discuss parts that were more difficult to predict.

Lost in Translation

This book is translated to English from Japanese, and as a result, the teacher should guide students in considering how translations might impact meaning. Through this examination, students will build background on the book's point of view. The teacher can accomplish this by asking questions about how translated texts might lose some original meaning and ask students to share any examples they know where the meaning of a word in the original language

What did we learn about the narrator?	What do we know about the narrator's gender?
• Likes math • Knows two people named Tetra and Miruka. Met them in high school • Thinks of math as a mystery. Characterizes it with words like "drama", "battle", "thrilling game", "competition of intellect", a "hefty weapon" (p. 1-2) • Strong memories of high school	• Nothing is revealed about the narrator's gender in this excerpt. • Narrator uses "I" pronoun. • Math is like a weapon/sword: Is this a masculine comparison? • Has feelings for Tetra and Miruka, but there seems to be some tension: does this relate to gender?
What do you predict about the narrator?	**What do you think the book will be about?**
• I really cannot predict that narrator's gender based on the prologue, but I wonder if the feelings for Tetra and Miruka and descriptions of these characters will lead to deeper relationships. • I predict that readers will discover why the narrator loves math so much. • I predict that the experiences this narrator has with math and these characters pushes the narrator to grow emotionally.	• I predict that a strong friendship develops between these three characters. • I predict the main character is very good at math. • I predict that it will be about a romance between Tetra and Miruka or Tetra and the narrator or Miruka and the narrator—or maybe a love triangle. The fact that the narrator has feelings for both of these other characters implies some tension there.

Figure 8.1 Prediction Chart. *Note*: This table was completed based on pages 1 to 2 in the text

is lost when it is translated or adopted by another language. Some examples include the following: *schadenfreude* (German), *sobremesa* (Spanish), *hygge* (Danish), *kummerspeck* (German), *akihi* (Hawaiian), ᐃᒃᑦᓱᐊᖅᐴᒃ (Iktsuarpok) (Inuit), Разлюбить (Razliubiti) (Russian), or 捨不得 (Shě-bu-dé) (Chinese) (Chueng, 2020). Each of these words does not have an English equivalent. For example, kummerspeck is a German noun that literally means "grief bacon," but this untranslatable word actually means gaining excess weight as a result of emotional eating. There might also be words that are lost in translation due to dialects within a language. When watching a British television series, for example, a character might feel "poorly," but a U.S. English speaker would say, "I am feeling sick." U.S. English speakers would use the term "poor" or "poorly" to describe something done badly or incompetently or even to describe a person in poverty.

The teacher could ask students to brainstorm terms that mean something different depending on the translation or audience. Discussion that follows can explore how meanings of words change when they are translated and whether a book or movie in its original language is the equivalent of its translation. The teacher could ask, "Since texts cannot be translated literally into another language, in what ways might bias impact translation (i.e., political, cultural, ideological, or economic beliefs or restraints)?" Students could work together to come up with examples of when bias impacts the way something was interpreted. Students might be able to make personal connections by sharing examples of when a slang word might take on a new meaning in different contexts (i.e., sick, cool, cold) and compare this to when words are translated incorrectly. When somebody says, "That's sick," for example, they could mean awesome or gross, depending on the context. If this was translated into a different language, the meaning might be lost in translation.

The teacher could have students visually indicate their answers to the following questions (i.e., raising their hands, thumbs up/thumbs down): *Do you trust what people say all the time? Do you trust strangers? Do you believe everything you read?* This would lead students to consider the more complex question: *When a text is translated, and the new audience is unfamiliar with the original text or language, how might a reader determine if the translation is valid?* This question centers on the trustworthiness of the translator, which helps to set the stage for reading and interpreting this book.

From there, ask students to look at the first sentence/character in *Math Girls*, which immediately identifies the gender of the narrator in Japanese language. The sentence, literally translated as "I can never forget," uses the first-person pronoun 僕 (*boku*), which is overwhelmingly used by males speaking to an unknown audience. Readers of the Japanese version would know the gender of the narrator immediately; on the other hand, the English version of the text doesn't mention the narrator's gender until the fourth chapter. The teacher could go on to ask students, "Is the translator reliable—since the original Japanese text reveals the main character is male, yet in the English version we don't find out that the character is male until the fourth chapter?" When thinking about the reliability of the translation, students have the opportunity to think critically about the way language and culture influence how we read a text, specifically this text, and the teacher could ask, "In what ways might knowledge of the narrator's gender change how we read and interpret the book?"

Did You Know?

Math Girls takes place in a Japanese school and features three high school–aged Japanese characters. To build background information that engages students with Japanese cultural elements that impact the text, a "Did you know?" activity to explore Japanese school culture is beneficial. The teacher can share "Did you know?" statements about Japanese school culture and ask students to write comparative statements that describe U.S. schools (see Table 8.1). By examining the differences between American and Japanese schools, students build background information that will help them to understand the school setting and culture in the book. As students write their comparisons,

Table 8.1 *"Did You Know?" Note:* Facts about Japanese and American schools were adapted from personal experience and knowledge, as well as from Freeman (2015), Ishikida (2005), and Mandrapa (2015)

Did you know?	Compare this fact to the U.S.
Japanese schools have a strict dress code. Most uniforms include some kind of tie, and also regulate students' shoes and backpacks, as well as clothes.	Some U.S. schools have a dress code (20%), but it is not as universal as in Japan.
Japanese schools do not have custodians. The students and teachers help keep the building clean.	In the U.S., students don't often care to pick up after themselves because they know somebody else will do it later. Schools are often really messy.
Japanese schools don't use substitute teachers. Students are expected to study quietly and independently.	*In the U.S. if a teacher is sick, a substitute teacher is called. Students would never be left unsupervised.*
Club activities are popular and serious.	Clubs and sports are popular with many students in the U.S. Some students are serious about them.
Students eat in their classrooms, and everybody eats the same lunch.	*Students bring lunches or buy lunch at school. Eating takes place in the cafeteria.*
Students can't fail a grade.	Students in the U.S. can fail grades and courses.
Japanese school year starts in April and ends in March.	U.S. schools start in the fall (August/September) and end in spring (May/June).
25% of high school students in Japan attend private schools.	*10% of U.S. students attend private high schools, much less than Japan.*
Japanese high schoolers must take entrance examinations to get into the best high schools in their district.	U.S. high schoolers take yearly tests, but these tests do not determine which high school a student is able to attend.
It is very competitive to get into a good middle school and high school in Japan.	It is not competitive to get into public schools in the U.S.; however, there are some places in the U.S. where it is competitive to get into charter or private high schools.
More than half of Japanese students attend "juku" or test preparation classes every day after school for several hours. These classes are expensive and parents spend thousands of dollars a year.	Some high school students take SAT or ACT preparation courses, but it's not a norm.
The federal government in Japan oversees the schools.	In the U.S., the states and school boards are in charge of the public schools.

the teacher can have students share in partners or as a whole class to compare answers and discuss differences that are interesting.

WHILE READING *MATH GIRLS*

In order to engage students with actively reading the text, we suggest activities that push students to think about the mathematical concepts critically and to consider the various ways characters in the story solve each problem and think about mathematics. During each scene, the narrator, Miruka, and Tetra each solve problems in different ways, and these varying approaches to each problem promote critical thinking about the process used in each situation. Tracking character arcs throughout the story provides students an opportunity to analyze each characters' attitude, mindset toward math, and other character changes throughout the story. This also provides students with an opportunity to explore their own attitudes toward mathematics and make connections between their own experiences and those of the characters.

Math Puzzles

The math problems presented in *Math Girls* create a number of opportunities for students to engage with mathematics. At the beginning of each chapter, Yuki's characters are given a math puzzle, which they work together to solve by the end of the chapter. The teacher can recreate this experience for students. Students can work problems before reading a particular section to help understand the mathematics of that section, or the teacher can create extension problems for students that use the same techniques as the book in order to further understand what the chapter is about. Here we provide two example activities, an example of an after-reading activity for Chapter 1, and an example of a before-reading activity for Chapter 4. In addition, we have also provided some summaries of key themes for Chapters 5 and 7 that could be used to plan or choose additional activities.

Chapter 1: Sequences and Patterns

In Chapter 1, readers meet Miruka and encounter some ideas at the core of mathematical philosophy. Miruka introduces herself to the narrator using a sequence "quiz," in which she gives the narrator a sequence of numbers such as $\{1, 1, 2, 3, \ldots\}$ (p. 4) and the narrator identifies a rule that fits the sequence (p. 6). Although the narrator finds a pattern and a rule in these sequences, Miruka says that these sequence problems have no right answers (p. 9), introducing a sequence that begins $\{1, 2, 3, 4, \ldots\}$ but continues with $\{6, 9, 8, 12, 18, 27, \ldots\}$ instead of continuing with $\{5, 6, 7, 8, 9, \ldots\}$. In this way, Miruka subtly makes a commentary on the nature of mathematical reasoning. Fundamentally, the study of mathematics is the study of what is possible and what is impossible. It is possible that a sequence that begins with $\{1, 2, 3, 4, \ldots\}$ continues $\{5, 6, 7, 8, 9, \ldots\}$ but it is also possible that the sequence continues $\{6, 9, 8, 12, 18, 27, \ldots\}$. As Miruka says, any number could follow 4 and the sequence would still be a valid sequence. This is because the sequence would still meet the definition of sequence. While students often believe that functions and sequences need to have formulas and rules, this is not true according to mathematical definition. Any set of ordered pairs where the initial element is not repeated is a function, and any function on the natural numbers is a sequence, whether there is a formula for that sequence or not.

But when would such a situation of a counterintuitive sequence arise naturally? Stylianides and Stylianides (2009) give us an example with the problem of cutting a circle.

Activity 1: The Circle and Spots Problem. Place different numbers of points around a circle and join each pair of points by straight lines. Explore a possible relation between the number of points and the greatest number of nonoverlapping regions into which the circle can be divided by this means (Stylianides & Stylianides, 2009, p. 329). Exploring the first few cases (see figure 8.2), we see the sequence is {1, 2, 4, 8, 16, . . .} and it appears that the circles follow the rule $S_n = 2^{n-1}$. But, if we continue to make circles, we see this pattern is broken. N = 6 gives us 31 regions instead of the expected 32, and N=7 gives 57 regions instead of the expected 64. When using this in the classroom, students should be allowed to place dots and count regions any way they like. Students will find that some configurations of dots will give fewer regions, but the teacher can record the maximum for each case. Students will predict that 32 regions in the N = 6 case are possible, even if they can't find the solution.

There are two key features to explaining why the correct number of regions for N = 6 is 31 and not 32. After students explore circles and make their predictions, the teacher should lead the students to focus on these two key features (Noy, 1996). The first is that every time we add a new spot to the circle, we also add N–1 chords (figure 8.2, dashed lines). So when adding the fifth spot, we add three chords. For the sixth spot, we add five segments. The second key feature is that the number of regions a chord creates is based on the number of existing segments the chord crosses. If a new chord crosses zero segments, it crosses only one region, cutting that region into two parts. If a new chord crosses three segments, it crosses four regions, cutting those four regions into eight. So every new chord adds new regions equal to one + the number of intersections (see figure 8.2).

Exploring the pattern of adding chords, students should create a table to see that the Case N = 4 adds three new chords with a pattern of 0 intersections, 1 intersection, 0 intersections. This gives a pattern of + 1 + 2 + 1 new regions. Similarly, in the case of N = 5, the new point adds four chords with a pattern of 0, 2, 2, 0 intersections for + 1 + 3 + 3 + 1 new regions, and for N=6, the new point adds five chords in a pattern of 0, 3, 4, 3, 0 for + 1 + 4 + 5 + 4 + 1 new regions, and 16 + 1 + 4 + 5 + 4 + 1 = 31. Students can then conjecture and verify that the pattern for N = 7 is + 1 + 5 + 7 + 7 + 5 + 1, giving 57 regions for N = 7.

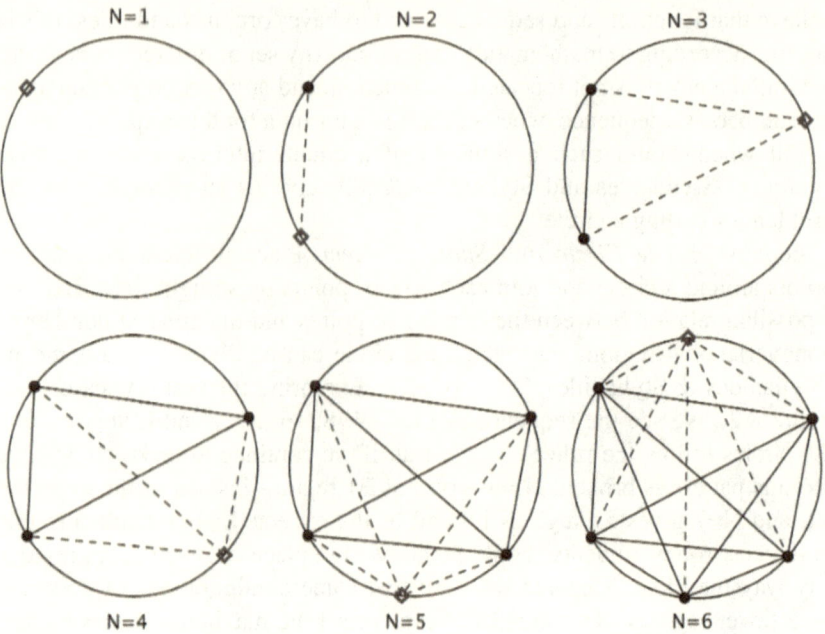

Figure 8.2 **The First Five Cases for the Circle and Spots Activity.** *Source*: Created by authors

After the activity, the teacher should organize a discussion about what students learned from the activity. The teacher should collect observations from students and record them on the board. Highlight student observations that advance the conversation toward two key learning goals: first, that as Miruka points out in this chapter, patterns we find in mathematics cannot be trusted to continue, and second, that we can understand a pattern in mathematics by carefully studying the steps we used to create that pattern. This second idea, of talking about and drawing conclusions from the actions we take in mathematics, is critical to understanding where mathematical rules come from, and is a problem-solving technique that the characters use repeatedly in later chapters. For example, in Chapter 2, Miruka explicitly asks the narrator to describe his steps (p. 14).

Chapter 4: Generating Functions

Chapter 4 introduces the technique of using a generating function to solve problems with recurrence relations. The core idea is that we can represent a sequence using a function in such a way that any arithmetic we do to the function, we do the same arithmetic to the sequence. For example, on pages

52 and 53, Miruka represents the sequence $\{1, 1, 1, 1, \ldots\}$ with the function $F(x) = 1 + 1x + 1x^2 + 1x + \ldots = 1/1 - x$. But if we multiply the equation by 2, we see that the function $G(x) = 2/1 - x = 2 + 2x + 2x + 2x^3 + \ldots$ represents the sequence $\{2, 2, 2, 2, \ldots\}$. Miruka uses this idea to split the generating function for the Fibonacci sequence $F(x) = x/1 - x - x^2$ into a sum of two functions $F(x) = R/1 - rx + S/1 - sx$ so that she can find the sequences corresponding to those functions and add them (pp. 54–65). Because of the irrational numbers involved, Miruka's solution to the Fibonacci sequence problem is quite complex, and since the technique of generating functions is also used in later chapters, we provide a simpler sequence of the problems based on Levin (2016) that illustrates the same strategy.

Activity 2: Finding a Closed Form with a Generating Function. These three problems are best used as an example in class, where the teacher provides a guided demonstration of the generating function technique midway through reading Chapter 4, with the goal of preparing students to read and follow Miruka's solution to the Fibonacci problem. After the class reads the first half of Chapter 4 (up to page 54), the teacher should discuss the problems below in a whole-class setting.

First demonstrate problem 1 in its entirety, and ask students to talk about and identify the steps. Then demonstrate problem 2, while increasing student participation by asking students what they notice, what ideas they have to do next, or what steps the class followed in problem 1. Problem 3 will require the teacher to set up the fraction decomposition, but ask students to find the values of A and B, and identify the relationship between the fraction terms, the solutions to problems 1 and 2, and Miruka's solution to $\{1, 1, 1, 1, \ldots\}$ (pp. 52–53).

Problem 1. Let S_n be the sequence defined by the recurrence relation $S_0 = 1$, $S_{n+1} = 2S_n$. Find the closed form of the generating function $S(x)$ for this sequence.

Solution. The first few terms of the sequence are $\{1, 2, 4, 8, 16, 32, 64, \ldots\}$ So the first few terms for the generating function $S(x)$ are $S(x) = 1 + 2x + 4x^2 + 8x^3 + 16x^4 + 32x^5 + 64x^6 + \ldots$

The recurrence relation tells us that the coefficient $S_{n+1} = 2S_n$ and the polynomial term for S_{n+1} is $S_{n+1}x^{n+1}$. So by substitution, we have $S_{n+1}x^{n+1} = 2S_n x^{n+1}$, and relating this to the polynomial term for S_n, we have $S_{n+1}x^{n+1} = 2xS_n x^n$. Using this, we can multiply $S(x)$ by $2x$ to generate a polynomial that will cancel nicely.

$$S(x) = 1 + 2x + 4x^2 + 8x^3 + 16x^4 + 32x^5 + 64x^6 + \ldots$$

$$2xS(x) = 2x + 4x^2 + 8x^3 + 16x^4 + 32x^5 + 64x^6 + \ldots$$

Subtracting these gives $(1-2x)S(x) = 1$, so the closed form is $S(x) = 1/1-2x$.

Problem 2. Let S_n be the sequence defined by the recurrence relation $S_0 = 1$, $S_1 = 3$, $S_{n+1} = 3S_n - 2S_{n-1}$. Find the closed form of the generating function $S(x)$ for this sequence.

Solution. Ask students to reconstruct and follow the steps from problem 1. The first few terms of the sequence are $\{1, 3, 7, 15, 31, 63, \ldots\}$ So the first few terms for the generating function $S(x)$ are $S(x) = 1 + 3x + 7x^2 + 15x^3 + 31x^4 + 63x^5 + \ldots$ Using the recurrence relation we have that $S_{n+1} = 3S_n - 2S_{n-1}$, and the polynomial term for S_{n+1} is $S_{n+1}x^{n+1}$. Substitution gives us $S_{n+1}x^{n+1} = (3S_n - 2S_{n-1})x^{x+1} = 3xS_n x^n - 2x^2 S_{n-1}x^{n-1}$. This suggests that we should multiply by $-3x$ and $2x^2$ to cancel nicely. We choose the negations so that we can add instead of subtract.

$$S(x) = 1 + 3x + 7x^2 + 15x^3 + 31x^4 + 63x^5 + \ldots$$

$$-3xS(x) = 0 - 3x - 9x^2 - 21x^3 - 45x^4 - 93x^5 + \ldots$$

$$+2x^2 S(x) = 0 + 0 + 2x^2 + 6x^3 + 14x^4 + 30x^5 + \ldots$$

Adding the equations gives $(1 - 3x + 2x^2)S(x) = 1$, so the generating function is $S(x) = 1/1 - 3x + 2x^2$.

Problem 3. Use the closed form generating sequence from problems 1 and 2 to find the closed form for the nth term of S.

Solution. Begin with a partial fraction decomposition of the generating function:

$$S(x) = \frac{1}{1 - 3x + 2x^2} = \frac{1}{(1-x)(1-2x)} = \frac{A}{1-x} + \frac{B}{1-2x} = \frac{A(1-x) + B(-1x)}{(1-x)(1-2x)}, \text{then}$$

$$1 = A(1-2x) + B(1-x).$$

Ask students to suggest values for A and B. When $x = 1$, $1 = -A$, so $A = -1$. When $x = 1/2$, $1 = B/2$, so $B = 2$. Our decomposition is $S(x) = -1/1 - x + 2/1 - 2x$. Next ask students to look back at the solutions to problems 1 and 2, and identify a relationship between those solutions and the terms of our rewritten $S(x)$.

We see that $-1/1 - x = (-1)1/1 - x$. Because Miruka found that $1/1 - x$ is the generating function for $\{1, 1, 1, 1, 1, \ldots\}$, we know that $-1/1 - x$ is the generating function for $\{-1, -1, -1, -1, -1, \ldots\}$ or the sequence $A_x = -1$. Similarly, $2/1 - 2x = (2)1/1 - 2x$. $1/1 - 2x$ is the generating function from problem 1, for the sequence $\{1, 2, 4, 8, 16, 32, 64 \ldots\}$, so $(2)1/1-2x$ is the generating function for $\{2, 3, 16, 32, 64 \ldots\}$, which is $B_n = 2^{n+1}$. Using the property that the generating function for a sum of two sequences is the sum of their generating functions, we have that $S(x) = -1/1-x + 2/1-2x$, so $S_n = A_n + B_n = -1 + 2^{n+1}$.

Chapter 5: Geometric Mean

In Chapter 5, Tetra and the narrator work on deriving the arithmetic-geometric mean inequality. The mathematics in this chapter is very accessible to secondary students, using only the difference of squares and rules of exponents. The narrator talks about playing around with formulas for the pleasure of playing with mathematics (p. 70); however, not all students are motivated by pure mathematics. Students may be interested in the applications of the geometric mean to provide some context for why this inequality is significant.

Here we would recommend that the teacher introduce problems that focus on exploring different applications of different kinds of means (arithmetic, geometric, and harmonic). Short problems where students work the problem and then discuss their methods as a class are ideal. While students discuss their work, time should be dedicated to developing a meaning of "average" or "mean" that emphasizes finding an equivalent constant action. So an average speed is the constant speed needed to travel the same total distance in the same total time, and an average interest rate is the constant interest rate needed to earn the same total amount of money in the same total time.

After discussing their work, the teacher can refer students back to the test by asking students about which activity (the Narrator's pure math activity or the classroom applied math activity) students thought was more interesting, and why. The teacher should encourage a variety of opinions here, with the lesson that there is no one right way to enjoy mathematics. Different mathematicians engage in and enjoy mathematical activity for many different reasons.

Chapter 7: Binomials

In Chapter 7, Tetra and the narrator work on the Binomial theorem, while Miruka and the narrator work on Catalan numbers. Both undertakings involve a dip into combinatorics. The Catalan numbers turn up in a broad variety of situations, such as the number of ways to correctly match n parentheses (p. 102) or the number of monotonic lattice paths that do not cross a diagonal (pp. 123–124), and a teacher may wish to explore some of the other applications of Catalan numbers. However, this chapter spends much less time on binomial coefficients. Binomial coefficients are more accessible and are fundamental to a broad variety of mathematical results, so we recommend activities that focus on some of the most famous connections for binomial coefficients: Pascal's triangle and the binomial distribution. By engaging with binomial coefficients using multiple representations and applications, students can have a similar experience to Miruka's experience with the Catalan numbers, but with more accessible mathematics.

Summary

Math Girls uses mathematical problems to drive the narrative of the story. A teacher can use and build on the problems of *Math Girls* in a number of different ways. The example activity for Chapter 1 on Sequences is all about showing an example of Miruka's claim that the pattern of a sequence can be counterintuitive (p. 9). The activity for Chapter 4 on generating functions is all about giving students an example midway through the chapter that will help them understand the new and powerful techniques that the characters are about to use.

In studying a chapter to create an activity, teachers should focus on identifying a key theme or attitude toward mathematics that the characters hold, and build on that perspective. Chapter 1 is about how mathematics is about studying rules and actions more than studying patterns (p. 9). Chapter 4 is about how we can use a tool kit from one area of mathematics (polynomial functions) to draw conclusions about a different area (sequences) (p. 55). Chapter 5 is about how mathematics can be fun (p. 70) and creates an opportunity to discuss the different ways that people enjoy mathematics or the different reasons that people have to do mathematics. Chapter 7 is about how a mathematics problem can be represented in many different ways, and how changing the representation can give new insights (pp. 122–123).

Across the book as a whole, there are consistent themes that the teacher can focus on to find the theme of each chapter. Themes that mathematics is about the study of rules and actions, that math rules make sense and fit together to create a puzzle game to play, that mathematics problems can be represented in many different ways, and that different people engage in mathematics for different reasons. An example of the last theme is how the characters of the narrator, Miruka, and Tetra develop in their attitudes toward mathematics and each other over the course of the novel.

Character Arcs

Though students will be studying mathematical concepts while reading each chapter, they can also examine characterization in order to connect and engage with the characters in the story. The main characters—the narrator, Miruka, and Tetra—are dynamic characters that change and evolve as the story progresses. Each has unique characteristics that define their

Table 8.2 Tracking the Characters

Character(s)	Event (include pg. number(s))	Character Trait(s)	How would you have responded in this situation? Would your response have been similar to the characters in the text or different?
Example 1 Miruka, narrator, and Tetra	(p. 141) Miruka finds the narrator and Tetra in the library, pats Tetra on the head and stomps on narrator's foot.	Miruka jealous? Tetra intimidated by Miruka-offers to leave. Narrator confused by the girls' responses.	If I was in this situation, I don't think I would have been jealous. I would have realized that it's a public space and school related activity. I may not have been doing math in my free time.
Example 2 Tetra and narrator	(p. 80–84) Tetra and the narrator talk about how both feel about math, Miruka, and each other.	Tetra: self-doubt. Tetra assumption = math easy for narrator, but narrator reveals math is hard. Narrator points out Tetra's strengths: language, his own strengths: technology, and Miruka's: math.	I have felt similar when I have been around people who are really good at certain things that I am not good at. In this situation, I may have reacted similarly to Tetra—not as confident toward the things that I struggle with.
Example 3 Narrator and Tetra	(p. 180–182) Tetra and the narrator attend the planetarium.	The narrator starts to question his feelings toward Tetra. He realizes that Tetra is asleep.	I have definitely been in a situation where I have considered words that I want to say to somebody and ran through all of the scenarios prior to speaking, only to lose the chance to say it. In a situation like this, I probably would have said something sooner than the narrator.
Example 4 Miruka, Tetra, and narrator	(p. 207–208) Tetra comes into the narrator and Miruka's classroom as Miruka solves a problem.	Tetra amazed by Miruka's equations on the blackboard. Miruka-kind: resets her lecture to include Tetra. Narrator realizes something has changed between the girls-Tetra and Miruka now friends?	I have definitely been in a situation where I've underestimated somebody and needed to change my approach when dealing with them. I think I would have reacted similarly as Miruka once I realized my mistake. If I was the narrator, I would have been surprised by the changed relationship between Tetra and Miruka.
Example 5 narrator	(pp. 116–121) Narrator ruminates over the problem he has been given by Mr. Muraki.	Narrator-reflective: looks at notes and applies what he learned from Miruka. In love? thinks about Tetra several times (i.e. "I should show this to Tetra", p. 119). Proud: Excitement when he solves pieces of the problem (i.e. "Of course, a sum of products", p. 118).	In this scene, readers can almost see the lightbulb going off in the narrator's head, and his rambling thoughts as he works on the problem are very realistic. I relate this to times when I have been writing a paper or working through a problem and suddenly I see the way to solve it. The excitement the narrator feels in his success is similar to how I would have reacted.

personalities and interest in mathematics, and readers can track changes to their attitudes and/or mindsets toward mathematics and changes in their feelings toward one another throughout the story. Since this story is told from a first-person narrator perspective, some of the typical characterization strategies, such as analyzing character's thoughts and feelings, might be limited, but students will be able to analyze character traits through dialogue between the characters, the narrator's thoughts and feelings about each character (and himself), and the reactions of others.

Students can track each of the main characters by using a graphic organizer (see Table 8.2). This approach can help them to focus on events that reveal attitudes and/or mindsets toward mathematics and changes in feelings toward one another. For example, to begin with, Miruka and Tetra feel tension whenever they are together with the narrator (pp. 49–50, 98–99, 105, 130, 140–141, 207). However, Miruka's attitude toward Tetra changes once she discovers Tetra's seriousness about learning math (pp. 212–217). During this scene, Miruka encounters the narrator and Tetra working on a math problem. This is when Miruka notices Tetra's notebook, "Hmm ... [. . .] You were teaching her the Taylor series for sin χ" (p. 212). She sees Tetra's detailed notes and attempts to solve the problem in the notebook pages, and she begins to teach Tetra the equation. Another example would be the main character and Miruka's interactions. When the narrator attempted to solve a problem (pp. 21–24), he shared his solution with Miruka, and her answer, "Well, it's right . . . but it's kind of a mess" (p. 24), could have frustrated him. But, he decided to ask Miruka, "Is there some way to make it simpler?" (p. 24). Throughout the text, the narrator often felt "juvenile and clumsy" compared to Miruka's "smart and elegant" style (p. 205). But, his interest in improving his math skills pushes him forward, and he welcomes the challenge.

AFTER READING *MATH GIRLS*

The following after-reading activities push students to think deeply about the attitudes toward mathematics portrayed in the book and to build upon previous activities they completed before and during their reading of the text.

Revising Attitudes toward Mathematics

After reading, students could repeat the quick-write activity. Students can write a second response to the prompt: *What does it mean to be good at math?* This response should be completed without referring to their previous quick-write, completed prior to reading. After students have been provided ample time to write their second response, ask them to take out their original response and compare what they wrote prior to reading *Math Girls* to what they wrote after reading the novel. To track changes, students can underline or highlight differences, marking the text to indicate any changes in attitude.

After comparing their individual responses, students can meet with a small group to discuss how their attitude changes relate to events in their experience reading the book. Students can refer to their "Character Arc" activity, using the "Tracking the Characters" graphic organizer as well as their annotated quick-writes to guide the discussion. The teacher can follow-up with small groups to provide students the opportunity to share any changes they noticed.

The teacher might follow-up the discussion with questions related to students' perceptions of how this experience impacted their own attitudes toward learning and engaging with mathematics. Some questions to consider include the following:

- How can the book's characters serve as models for how we should think about math?
- How could the character's challenges with math help you to think about the stresses of math in a constructive way?
- Seeing how both Tetra and the narrator work through difficult math problems despite their lack of confidence, how can they serve as a model for you as you work through difficult problems?
- Why do you think these characters find math so compelling as a hobby?

Knowing that characters in the text view mathematics as difficult and challenging, yet engage with it enthusiastically, might create an opportunity for students to discuss math anxiety, self-efficacy, perseverance, and grit.

A Mathematical Society

Students could replicate the mathematical society that Yuki creates in the world of the book. They will create math problems for classmates to explore. For example, problems are presented in *Math Girls* in shaded boxes throughout the text, and various characters pose and solve the problems (pp. 4, 5, 6, 14, 21, 37, 38, 43, 56, 89, 93, 95, 102, 109, 151, 155, 160, 164, 186, 187, 209, 221–222, 235, 237, 242). Just as the characters in Yuki's book, students can design a problem that is specifically meant to be challenging, interesting, complex, and solvable for the students in their own math class. Audience is an important factor in this process; recall that Mr. Muraki created specific problems for each student (pp. 155, 186, 209, 221–222), based on their individual ability levels.

Therefore, students will need to tailor tasks based on knowledge of their peers' interests and the curricula already addressed in their course. Students can consider whether this problem would be a review, would follow the current curriculum, or would challenge students with new content. Each problem should include a clear description, be appropriate, and provide room for exploration. The problem should be simple to understand, and the description of the problem should be short and clear. The problem should ask students to complete techniques they are familiar with or can figure out based on what they already know. There should be multiple paths to the solution and multiple ways to represent the problem/solution. Additionally, the problem should integrate a unique combination of math skills.

Students could solve their problem ahead of time before passing it on to a classmate, and the teacher should ensure that solutions and problems make sense and are correctly worded and solved. Solutions should demonstrate an understanding, not just of the problem itself, but also of the way that the problem connects to the topics and problems presented in *Math Girls* or to other areas of mathematics. In addition to providing a problem for a classmate, each student could also receive, and solve, a problem that has been provided to them by somebody else. They can solve the problem and turn that solution into the teacher and the other student.

Ask students to turn in a copy of their original problem, their own solution, a description of what makes the problem challenging and interesting, their classmate's solution, and a reflection on the differences between their solution and their classmate's solution, including what the student learned by giving the problem to someone else.

Revisiting Characterization

The teacher can provide specific prompts to students as they analyze the characterization they uncovered during the "character arcs" activity in order to focus specifically on traits related to learning and engaging with mathematics. By using these prompts, the teacher can focus students' analysis on specific math-related changes to each character. Sample prompts include the following:

- How do characters in the story respond to challenges (for example, thinking about the math concept/problem, pp. 6–8, 116–121; asking for help, pp. 15, 68–73)?
- Why does Miruka's attitude toward Tetra to change (pp. 212–217)? How does Tetra's interest in math influence this (p. 212)?

Exploring Math, Culture and Stories through Math Girls   149

- How does skill with math impact the characters' perception of themselves and others (pp. 80–84)?
- How do the characters feel about getting a problem from Mr. Muraki (pp. 101–102, 167–168, 209)? From another student (pp. 13–14, 43, 89, 93)?
- How do the characters interpret the act of asking for help? How does seeing other characters ask for help affect their perception of those characters (asking for help, pp. 15, 68–83; perceptions, pp. 21, 34, 65, 80–83, 162)?
- What about a particular problem would a character in the story identify as fascinating ("A chill ran down my spine," p. 44; "Miruka this is incredible," p. 98)?
- What about a particular solution would a character identify as beautiful (pp. 53–54; 127–129, 180, 216)?
- In what ways does the main character persevere in the face of setbacks, frustrations, or disappointments (pp. 82–83, 165–171, 205–207, 231–232, 251)?
- In what ways does the main character keep trying to solve a problem in fewer steps or simpler terms—even when he has already found the solution (pp. 24–26, 98, 117–121, 165–171)?
- What risks do the characters in the story take? Why do the characters take these risks (asking for help, pp. 15, 68–83; taking chances on problems, pp. 212–219)?
- Why does the main character continue to go to Miruka for help when he sometimes feels dumb compared to her (pp. 43, 57, 93, 205)?
- Why does Tetra turn to the narrator for help (pp. 33, 69, 202)?

BEYOND *MATH GIRLS*

There are several extension activities that build on the before-, during-, and after-reading activities presented in this chapter. These extension activities are recommended, as they move students into a deeper understanding of the concepts presented in the text.

But Wait, There's More

Yuki hints at much deeper mathematical problems than students might be ready for, and this creates an opportunity for extension and exploration. Toward the end of the book, there are several opportunities for the teacher to build on student interest. Yuki presents the problem of partition numbers in Chapter 10 (p. 221), as the final problem of the book. While solving the partition problem is extremely challenging, students might be interested in studying the history of the solution. Emory University (2011) provides a video titled *New Theories*

Reveal the Nature of Numbers on the history of partition numbers. In this video Ken Ono talks about what partition numbers are, gives a history of various mathematicians' attempts to find a formula for partition numbers starting from Euler in the 1700s, and talks about current work of finding patterns in the partition numbers. Bressoud and Propp (1999) also give a history of a related problem, the alternating sign matrix conjecture. In this short paper, Bressoud and Propp explain what an alternating sign matrix is, and how attempting to count the number of alternating sign matrices was eventually discovered to be the same problem as a three-dimensional version of the partition problem. Similar to Chapter 7 on the Catalan numbers, this paper shows how a single math problem can have several different representations.

Mathematical Careers

Students could be assigned a project to look into careers that use mathematics, or to choose or identify a career for the main characters of the book. The book primarily concerns itself with algebra and number theory puzzles, so related careers such as mathematics professor or cryptographer might be interesting to these characters. Students can also research the broad ranges of other professions that require significant mathematics, including physics, engineering, computer science, computer graphics, game design, finance, insurance, and economics. These professions also use mathematics in different ways. Some professions use mathematics more as a tool, while other professions see mathematics itself as a kind of puzzle or game as it was portrayed in the book. Students could identify not only the mathematics used in these professions, but also differences in what mathematics is used, how mathematics is used, and the attitudes and perspectives of professionals toward mathematics in these areas.

Modeling a Story

Since this book presents text that requires readers to code-switch between English and mathematical terms, writing a story that includes characters solving or working through math problems would require students to use academic language from mathematics, as well as apply literacy techniques. Students could use a graphic organizer to plan out the story (see figure 8.3). This graphic organizer might look different from traditional story maps since it would need to consider these code-switches, as well as determining math complexity and problems to include. Students could use the problem they created in the after-reading activity, "A Mathematical Society," as the basis for their story. Use the following story map to write a TWO-character story

Exploring Math, Culture and Stories through Math Girls   151

	Brainstorming the Story	
What is your math problem?	A ball is dropped down a board with rows of pegs. Every time the ball hits a peg, it gets knocked left or right before falling down to the next row and hitting another peg. Count the number of paths that the ball can take to each space on the board.	
What is the solution?	The solution is Pascal's triangle	
What are the steps of solving the problem?	1) At the top of the board, there is one path to the left and one path to the right. 2) In the middle of the board, a path to a space comes from the spaces diagonally above it. 3) The number of paths to a space is the sum of the paths to the spaces diagonally above it.	
Who will be the characters in your story?	Character 1 (expert) *Shelly*	Character 2 (person solving problem) *Carlos*
What will be the setting?	*When* Lunchtime	*Where* Coffee Shop with a peg board
	Mapping the Story	
Introduction: How will you introduce the story and/or math problem? • Where are the characters? • Who are the characters? • Why do they want to solve the math problem?	• Shelly and Carlos are friends having lunch at a coffee shop • They notice the coffee shop has a peg board where you can drop a ball to win a free coffee • They wonder how many paths there are to the coffee prize at the bottom	
Rising Action: Using the steps to solve the problem, write responses that each character might say/do at each of these steps. • Where might somebody struggle with this problem? Include at least two places where the characters in your story struggle and then show how the character overcomes.		Carlos notices that at the top of the board there is one path to the left and the right
		Carlos gets stuck trying to figure out a rule for the paths in the middle. There are too many of them!
	Shelly suggests relating one space to the row above it	
		Carlos figures out the sum property (step 3)
Climax: The characters solve the problem. • How do they respond? • Does the expert character respond like Miruka and provide feedback to make the solution more elegant? • Do they celebrate their success?		Carlos fills out the spaces on the board and notices that they make a triangle
	Shelly introduces the history of Pascal's triangle and points out some number sequences in the diagonals	
Resolution: Design an ending to the story. • How does the story end? • What actions do the characters take to plan another mathematical society "meeting"?		Carlos drops a ball, but does not win a free coffee
	Shelly points out that lunch is over and it's time to get back to work	
		Carlos suggests meeting at the same time next week

Figure 8.3 Sample Story Graphic Organizer. *Source*: Created by authors

where the characters solve a math problem. Have students take the characters through the steps of solving the problem, modeling after how author Hiroshi Yuki's characters solved problems in *Math Girls* (i.e., step by step, including mathematical expressions and mathematical discourse). Additionally, they could include at least two instances where the character makes an error and the "expert" character has to step in to help.

CONCLUSION

Math Girls described an environment where students challenge themselves with math problems for the pure joy of doing mathematics, making connections, and seeing solutions come together. This chapter presented several strategies for incorporating *Math Girls* into a high school mathematics curriculum. By modeling and discussing the struggles of the characters, and their attitudes toward struggle, students can discuss and investigate their own attitudes about challenge and perseverance in math. This contemporary novel—that incorporates mathematical equations and prose—helps readers to learn academic language in mathematics and pushes readers to consider attitudes and values that they hold toward math. Because it is a translated text, set in Japan, students also have the chance to explore culture and language in the activities presented in the chapter. By incorporating specific math skills that students are expected to learn and practice in core mathematics standards, as well as literacy skills, reading *Math Girls* can provide an opportunity for students to develop deeper understanding of both mathematics and literacy.

REFERENCES

Braden, L. S. (1991). My favorite rate problems. *The Mathematics Teacher, 84*(8), 635–638.

Bressoud, D., & Propp, J. (1999). How the alternating sign matrix conjecture was solved. *Notices of the AMS, 46*(6), 637–646.

Chueng, P. (2020). *20 untranslatable words with no English equivalent*. Swap Language. https://swaplanguage.com/blog/lost-in-translation-20-meaningful-words-with-no-english-equivalent/.

Emory University. (2011). *New theories reveal the nature of numbers* [Video]. https://www.youtube.com/watch?v=aj4FozCSg8g.

Freeman, E. (2015). *9 ways Japanese schools are different from American schools*. https://www.mentalfloss.com/article/64054/9-ways-japanese-schools-are-different-american-schools.

Hobbs, J. (2004). *Bridging the cultural divide: Lexical barriers to translation strategies in English translations of modern Japanese literature*. https://translationjournal.net/journal/28litera.htm.

Ishikida, M. Y. (2005). *Japanese education in the 21st century*. iUniverse, Inc.

Levin, O. (2016). Generating functions. In *Discrete mathematics: An open introduction*, (3rd ed.) (pp. 295–306). University of Northern Colorado.

Mandrapa, N. (2015). *Interesting facts about Japanese school system*. https://novakdjokovicfoundation.org/interesting-facts-about-japanese-school-system/.

Marzano, R. J., Pickering, D. J., & Pollock, J. E. (2001). *Classroom instruction that works: Research-based strategies for increasing student achievement.* Association for Supervision and Curriculum Development.

Noy, M. (1996). A short solution of a problem in combinatorial geometry. *Mathematics Magazine, 69*(1), 52–53.

Yuki, H. (2011). *Math girls.* Bento Books, Inc.

Zazkis, R. (2000). Using code-switching as a tool for learning mathematical language. *For the Learning of Mathematics, 20*(3), 38–43.

Chapter 9

Making a Deal with *The Number Devil*

Scaffolding, Deepening, and Extending Students' Mathematical Literacy

Amanda Huffman, Rachel Colby,
Jenna Repkin and Shelly Furuness

The Number Devil: A Mathematical Adventure (Enzensberger, 1997) has long been identified by mathematics educators as a worthy choice to incorporate into mathematics curricula from middle grades through high school Calculus. There is recognition among mathematics teachers that while *The Number Devil* may have potential points of frustration for mathematically minded readers for its "treatment of mathematics as trickery" and "the book's informal treatment of vocabulary associated with the mathematical topics presented," (Anthony, Kolodziej, & Meadows, 2017, p. 148) there is so much that can be used to engage learners in thinking mathematically and developing stronger mathematical literacy.

The Number Devil creates an engaging, contextualized, and fun way to learn about challenging concepts in mathematics through the dreams of Robert, a student who dislikes math. Through creative language such as making numbers hop, prima donnas, taking the rutabaga, vroom!, unreasonable numbers, coconuts, handshakes, and Bonacci numbers in order to explain concepts including square roots, triangle numbers, and the factorials—students can be engaged in a fun way to learn about challenging concepts in mathematics through the dreams of Robert. It is through the book's creative and accessible language describing mathematical concepts that teachers have an entry point to scaffold, deepen and extend their students' mathematical literacy and use of academic language.

THE NUMBER DEVIL: A MATHEMATICAL ADVENTURE BY HANS MAGNUS ENZENSBERGER

To Robert's dismay, the Number Devil, a mathematical whiz the size of a grasshopper, appears in his dreams. The Number Devil tries to win Robert over by showing new approaches and insights into the world of mathematics. Over the 12 dream-filled nights described in the novel, Robert and the Number Devil explore number sense concepts including a discussion of rabbits relating to Fibonacci numbers, the importance of the numbers one and zero, infinity, and the classifications of numbers. Robert develops a better understanding of these concepts in mathematics and even stuns his teacher with the correct answer to a problem in class.

CONNECTING MATHEMATICS AND LITERACY

The Principles and Standards for School Mathematics (National Council of Teachers of Mathematics [NCTM], 2001) outlines five process standards: problem-solving, reasoning & proof, communications, connections, and representations. Mathematical proficiency is laid out in the National Research Council's *Adding It Up: Helping Children Learn Mathematics* (Kilpatrick, Swafford, & Findell, 2001) and includes conceptual understanding, procedural fluency, strategic competence, adaptive reasoning, and productive disposition. The process standards and the development of mathematical proficiency are being fostered in mathematical thinkers alongside the three stages of mathematical literacy (NCTM, 2000): declarative, procedural, and conceptual. These three mathematical literacy skills are critical for students to develop and strengthen their mathematical understanding across all mathematical content areas. Utilizing narratives in mathematics classrooms and building before-, during-, and after-reading strategies into mathematical exploration helps to strengthen what effective readers do—"make text-to-self, text-to-text, and text-to-world connections that aid in their comprehension, retention, and future retrieval of the information in the text" (Keene & Zimmerman, 1997, as cited in Anthony, Kolodziej, & Meadows, 2017, p. 149.).

BEFORE READING *THE NUMBER DEVIL:* *A MATHEMATICAL ADVENTURE*

Assessing Prior Knowledge: The Devil Is in the Details

In order to gauge students' knowledge of key mathematical concepts before reading *The Number Devil*, a baseline assessment can be given, similar to the

Directions: Answer each question completely to the best of your knowledge. Be sure to show all of your work.

Number Sense:
1. Write the next two numbers of the following sets of numbers:
   a. Prime numbers: 1, 3, 5, 7, 11,
   b. Perfect squares: 1, 4, 9, _____
   c. Even numbers: 2, 4, 6,
   d. Odd numbers: 3, 5, 7, _____

2. Use the following number to answer the questions below: 147.369
   a. What number is in the tenths place?
   b. What number is in the hundreds place? _____
   c. What is the value of the number 7?
   d. What is the value of the number 6?

Algebra:
Solving the following:
1. $2^2 =$
2. $3^1 =$ _____
3. $3! =$
4. $\sqrt{16} =$

Geometry:
1. What is pi (π) approximately equal to?
2. Identify four different types of shapes.

Problem Solving:
1. When leaving a birthday party, everyone shakes everyone else's hand while saying goodbye. If there were 5 people at the party, how many handshakes were made?
2. If a class had 4 students and the 4 students sat in one row of desks. How many different seating arrangements could the teacher make so that no two seating arrangements were the same?
3. Write the next three terms in the following pattern:
   a. $\frac{1}{2}, \frac{1}{4}, \frac{1}{8}, \frac{1}{16}, \ldots$
   b. $\frac{1}{2}, \frac{1}{3}, \frac{1}{4}, \frac{1}{5}, \ldots$
   c. $\frac{1}{2}, \frac{1}{4}, \frac{1}{8}, \frac{1}{16}, \ldots$

Figure 9.1 Pre- and Post-Assessment. *Source*: Created by authors

one seen in figure 9.1. The pretest features content that the Number Devil explores with Robert in his dreams throughout the novel. The concepts in figure 9.1 are some of the key concepts the Number Devil discusses with Robert. The concepts provided encompass four areas of mathematics (number sense, algebra, geometry, and problem-solving) often found in all levels of mathematics. The pre-assessment can be altered to meet the level of your students as well. For instance, the section on geometry or the problem-solving questions involving combinations can be removed if students are not studying those topics yet. Additionally, content areas can be further elaborated if the teacher's intention is to focus more on specific concepts being discussed. This pre-assessment is intended for the teacher to gather information and should not penalize the student. The data gathered from the pre-assessment offers teachers insight into what the students may or may not already know or understand. This can lead to an adjustment to the amount of time spent on particular concepts, such as focusing time on correcting a misconception observed in the pretest.

Getting (Re)Acquainted

A vocabulary knowledge-rating chart is a tool for teachers to assess how much students understand about specific mathematics topics. Students rank their understanding of mathematical terminology as one of the following: can define it, use it, teach it; heard it, seen it; or do not know. Students then have space to provide definitions and examples to demonstrate their understanding or possible understanding. Teachers can then use the information in the chart to help guide students in reading and provide additional background and support of key vocabulary when needed. Two different vocabulary knowledge-rating charts, one with the traditional mathematical term and one with the Number Devil's terminology, can be utilized, or a merged document including terminology from the novel can be used. The merged document can be seen in figure 9.2. It is important to realize that while the Number Devil creates some of his own terminology with the use of visual metaphors throughout Robert's dreams, each term directly correlates to a specific mathematical concept and it is crucial for students to know the mathematical terminology to develop academic language necessary for advancing mathematical literacy.

Starting Mathematical Conversations

Chalktalks can be one good way to see what students already think and know about the mathematical concepts covered in *The Number Devil*. Constructivist learning theory assures us that learning does not occur in isolation; it is constructed through scaffolded social interactions (Vygotsky, 1978). In other words, learning happens when students are talking with each other and the teacher about the content. Teachers can shape the social interactions students have while simultaneously building academic language when they encourage students to talk about mathematics. A chalktalk can help set the stage for students to read the novel as they begin thinking about some of the mathematics they will encounter while reading, as well as to help gauge students' prior understanding of the academic language that will be introduced, such as Pascal's triangle, the Pythagorean Theorem, factorials, and square roots. Chalktalks can be conducted literally, as it sounds. Pose a question or write a phrase/statement/mathematical equation or expression on the board. Then provide each student with a piece of chalk, possibly even a different color of chalk. Give students a designated amount of time to write everything that comes to their mind in response to the prompt on the chalkboard. During this time, the students cannot talk to one another, and can only respond and communicate by using their piece of chalk (i.e. writing words, drawing pictures, circling, starring, etc.). If a chalkboard or chalk is unavailable, this activity can easily be adapted for use with a dry erase board or with poster paper and markers.

This activity can also effectively incorporate purposeful movement by asking students to begin the chalktalk "conversations" at one location on the board or classroom and then to physically rotate to another location once they have contributed to the silent, chalktalk conversation. The teacher can monitor the chalktalk and direct students to add their thinking to underdeveloped conversations. Try different ways and see what invokes the most productive discourse. At the conclusion of the silent time, allow students to step back

Making a Deal with The Number Devil

Word	Definition	The Number Devil Page Number(s)
Combinations (*Broom brigade*)	The number of possible arrangements of objects where order does not matter	pp. 161–167
Dodecahedron (*Pentagon ball*)	A solid figure with twelve flat faces	pp. 206
Factorial (*Vroom!*)	The product of an integer and all of the positive integers less than it represented through an exclamation mark (i.e. 5! = (5)(4)(3)(2)(1)=120)	pp. 156–157, 179, 228
Fibonacci numbers (*Bonacci numbers*)	A sequence of numbers where the next term is the sum of the previous two terms (i.e. the Fibonacci numbers for n=1 would be 1, 1, 2, 3, 5, 8, 13, 21, 34, . . .)	pp. 108–121, 139–140
Irrational numbers (*Unreasonable numbers*)	A real number that cannot be written as a fraction	pp. 75, 78, 195
Natural numbers (*ordinary, garden-variety*)	Positive whole numbers, not including zero	pp. 55, 173–175
Octahedron (*double pyramid*)	A solid figure with eight flat faces	p. 205
Pascal's triangle (*Number triangle*)	A triangular array created through the summation of adjacent terms of the previous row	pp. 128–145
Permutations (*Changing places*)	The number of possible arrangements of objects where order does matter	pp. 149–157
Prime numbers (*prima-donna numbers*)	Numbers divisible by only 1 and itself	pp. 55–63, 176
Raising to a higher power (*Hopping*)	The product of a number and itself multiple times (i.e. 4³=4·4·4=64)	pp. 38–41
Square roots (*Rutabagas*)	The inverse of squaring. Given a number, try to figure out what number squared gives you this number	pp. 76–80
Squaring (*Hopping twice*)	Multiply a number by itself	pp. 79–80, 136, 179
Taking the root (*Hopping backward*)	Given a number, try to figure out what number squared, cubed, or any other power, gives you this number	p. 76
Taking the square root	Inverse operation of squaring	pp. 76–80
(*Taking the rutabaga*)		

Figure 9.2 *The Number Devil* **Vocabulary Knowledge-Rating Chart.** *Source*: Created by authors

and look at what has been put together. Then, after students have had time to take in everything from the chalktalk, lead a discussion for students to further express themselves, draw conclusions, and answer any lingering questions. Feel free to choose one main question or a handful of questions/topics to explore in the chalktalks. Possible questions, connected to specific content discussed in *The Number Devil*, to stimulate students' thinking before reading the novel include, but are not limited to, the following:

- What does a mathematician do? In what ways are you a mathematician (pp. 57–60, 62, 108, 224–226, 238–244, 246–247)?
- What is zero (pp. 32–36, 41, 245)?
- Pascal's triangle: What do you know about it (pp. 128–145)?
- What characteristics make a number pattern? Can you give some examples (pp. 174–176, 180–187)?
- Pythagorean Theorem: What do you know about it (pp. 81–83)?
- What are geometric shapes? What do you know about these shapes (pp. 203–210, 246)?
- What are prime and composite numbers (pp. 55–63, 156)?
- How many numbers are between 0.0 and 1.0 (pp. 17–20)?
- What are some different types of numbers? Or different classifications of numbers (pp. 34–45, 55–63, 75, 78, 93–103, 108–121, 128–145, 159–161, 173–176, 194–195, 242, 252–253)?
- What is a factorial and what is it used for (pp. 156–157, 179, 228)?
- If everyone shakes hands with everyone else in your group, then how many handshakes took place? What about if everyone included everyone in the class (pp. 158–161)?

Each of these questions posed in the chalktalk connects to one of the concepts explored within *The Number Devil*. The chalktalk serves as a way to access prior knowledge of students and these written conversations can be returned to again and again as students construct new and expanded understandings.

Feeding the Conversation

Another way to engage and immerse students in the Number Devil's dream world before they begin reading is to share a YouTube video clip called *Number Devil* (tennisdog2468, 2008) in which the first night's dream from the first chapter of the book is acted out through drawings. This video mimics and adds to the illustrations found throughout the novel (e.g. pp. 10, 13, 14, 17). The animation brings the novel alive for students, draws them in, and adds a visual and a voice for both the Number Devil and Robert. At the conclusion of the video, teachers can work with students to discuss their reactions, predictions, and connections to the clip. *What did they notice? What do they wonder? What questions do they have?*

Each of the strategies employed before or at the beginning of instruction can serve to inform the teacher of students' baseline knowledge of the key concepts addressed in the unit. Further, these strategies can increase student interest, connections, and confidence with the mathematical vocabulary and concepts.

WHILE READING *THE NUMBER DEVIL:* A MATHEMATICAL ADVENTURE

Thinking Mathematically throughout the Reading

After watching the video and initiating a discussion around the clip, begin the novel study by reading aloud the first chapter as a class. Rarely does a mathematics unit begin with a read aloud of a novel. This strategy is useful in simulating student engagement and activating the brain's attention by providing purposeful novelty in the presentation of the content. Reading the first chapter aloud also provides a chance for the teacher to model how students should approach reading this novel, including having a pencil nearby to make notes or do quick mathematics problems alongside Robert and the Number Devil. There are a few good stopping places for teachers to model thinking aloud. One place to stop is at the top of page 16 when Robert questions his belief in the Number Devil's statement that "there's an infinite number of numbers." Another place could be at the top of page 18 as the Number Devil introduces the chewing gum example. Here, the teacher could stop before it is completed in the book and have the students walk through the example as class with a piece of chewing gum. Yet another opportunity to model mathematical thinking is on page 24 where the teacher could challenge the students to find the answer to the problem Robert poses: $11,111,111,111 \times 11,111,111,111$. The teacher could also stop the students just before this page and have them work through the multiplication of similar patterns like 11×11; 111×111; $1,111 \times 1,111$; and so on. This strategy would allow students to see a pattern develop. They could then predict and explain the pattern of how each product is a palindrome for the problem the Number Devil is about to explain to Robert. In this way, students are activating and accessing their prior knowledge and mathematical skills in service of new and expanded mathematical literacy.

Dream Journal

The dream journal is an activity that aligns well with writing across the curriculum and helps to develop contextualized academic language. Students write about each chapter in the novel to show their comprehension of what they read. The students can be given the same prompts for each journal entry, such as the following:

- Using your own words, describe Robert's dream. As you write, pretend you are explaining the chapter to someone who has no knowledge of the concepts. Use both the mathematical language and the Number Devil's language to explain the dream.
- Describe something in the chapter that you had heard about before, but understand better now that the number devil has explained it. What did the Number Devil say that helped it make sense? If you were in Robert's place, what questions would you still have for the Number Devil?

Teachers also may choose to add additional prompts to each chapter based on its content. Here are a couple of prompts aimed at supporting students' rehearsal of academic language and review of the mathematical concepts following specific dreams:

- Following reading *The First Night* and *The Second Night*, have students journal about all the ways in which numbers can be expressed or written. Some guiding, or specific, questions could include the following: *How many ways can you express your age? How many ways can you express the year it is? How many ways can you express the number (give a specific number)?* This particular question invites space for differentiation in the task as some students may not be able to handle as big of a number as others.
- *The Third Night* includes discussion on prime numbers and includes the statement, "'Now tell me, dear boy, what the first prima-donna numbers are.'" (p. 55). The Number Devil invites the reader to continue narrowing the numbers between 2 and 50 so that only prima-donna (prime) numbers remain (p. 60). A replica of the Number Devil's prima-donna chart can be seen in figure 9.3. After students complete the Number Devil's task, pose the following statement: without the use of technology, or than a scientific calculator, you have [x] minutes to find the largest prime number that you can. Following the journaling time, have students share out their prime numbers and students can check to see if everyone has in fact come up with a prime number.
- In *The Fifth Night*, Robert learns about triangle numbers and quadrangle numbers. Have students hypothesize about pentagonal and hexagonal numbers in their dream journals.
- After completing *The Eighth Night*, students can spend time writing in their journal about how many handshakes would occur if everyone in their

2	3		5	7	
11	13			17	19
	23				29
31				37	
41	43			47	49

Figure 9.3 Chart of Prima-Donnas (Prime Numbers) between 2 and 50. *Source:* Created by authors

classroom shook one another's hands. In this exercise, answers will vary based on class size, but all students in the same class should arrive at the same answer, although some may or may not include the teacher in the count for the class. If there are 20 people in the class, there will be 190 handshakes in the classroom.
- The Number Devil teaches Robert about the concept of infinity in *The Ninth Night*. Students can write, in their own words, what they learned about infinity, whether that idea makes sense to them, and what questions they have.
- In *The Eleventh Night* the Number Devil and Robert talk about proving the various concepts they have discussed throughout this novel. Ask your students if there are any math concepts they have learned about that they do not understand why they work.
- *The Twelfth Night* brings the novel to a close. Students can write a final journal entry on their feelings of the novel. Did they like it? What did they like best? Did they learn anything new that has stuck with them throughout the novel?

In addition to supporting academic language use, the dream journals also serve as a way for the teacher and/or peers to identify misconceptions about mathematical concepts as students expand on their understanding from reading the chapters. Reading the journals before having a class discussion may help determine what misunderstandings students have from the chapter and what questions need to be addressed. However, if the classroom community is welcoming and students are comfortable sharing their findings/understandings, the discussion could take place immediately after students have been given the opportunity to journal. Either way, the discussions about each chapter can be based on student journal responses and questions they have. During these discussions, students have an opportunity to verbalize their journaling and their response to the prompts. Then their classmates have the chance to react to their responses, and the conversation between Robert and the Number Devil will evolve and come to life in your classroom among your students.

Activities Built into *The Number Devil*

The text itself offers some tasks for the reader to complete. These tasks encourage the reader to build on their understanding from the dialogue between Robert and the Number Devil. These mini-tasks can be utilized as independent, partner, group, or even whole-class work. Some tasks offer more exploration than others.

- At the conclusion of *The Third Night*, the Number Devil reveals one last trick about prima-donna numbers and challenges the reader to try to express 27 as the sum of prima-donna numbers (p. 64). (Answer: $17 + 7 + 3 = 27$).
- Toward the end of *The Fifth Night*, the Number Devil asks the reader if they continue to divide up the squares of squares, "What do you get if you add them together?" (p. 103). (Answer: $1 + 3 + 5 + 7 + 9 = 25$). This task reinforces the Number Devil's discussion on perfect squares.
- After the Number Devil's discussion on Bonacci numbers at the end of *The Sixth Night*, the Number Devil presents Bonacci numbers in a tree and asks the reader, "How many branches are there by the time you reach the top, line nine?" (p. 121). (Answer: 34).
- In *The Seventh Night*'s discussion of triangle numbers, the Number Devil guides the reader to discover the pattern if all the numbers that can be divided by four are highlighted in the triangle (p. 145). (Answer: See figure 9.4.).
- *The Tenth Night* focuses on geometric shapes. At the conclusion of the dream, the reader is guided in the creation of a geometric figure, "a ring with ten little pyramids" (p. 209). For this task, the net of the figure is provided on page 210 of the novel in case the Number Devil's directions are unclear.

Beyond the Think-Pair-Share: Collaborate, Connect, Create

The Number Devil often informs Robert of mathematical phenomena that Robert generally questions, or the Number Devil poses explicit questions for Robert while exploring the mathematical concepts. The questions from both the number devil and Robert are for the reader as well, not just Robert or the Number Devil, respectively. These questions create an opportunity to have students collaborate with their classmates about the mathematical concepts, connect the mathematics to the real-world, and create a representation of or situation involving the mathematics. Providing students with the opportunity to process what they are reading and make sense of it on their own helps build literacy skills, as well as

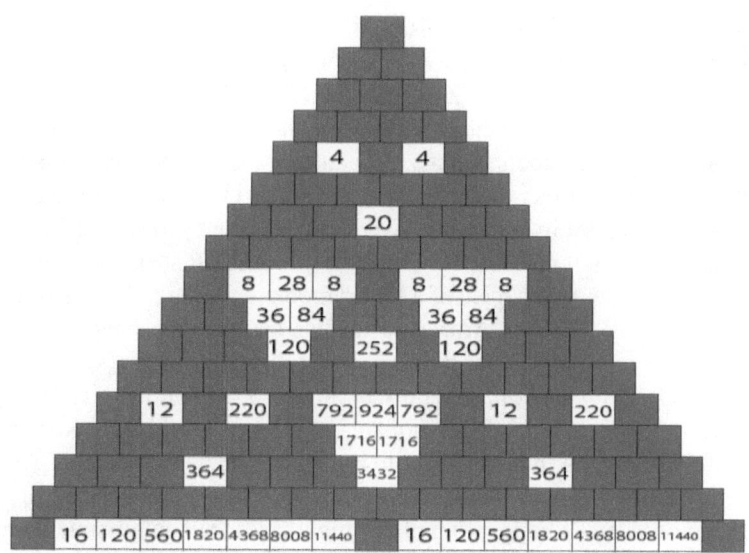

Figure 9.4 Pattern Created by Eliminating all Numbers Divisible by Four. *Source*: Created by authors

mathematical skills. Below is a list of some of the explicit questions raised by either the Number Devil or Robert providing the perfect opportunity for students to be collaborating, connecting, and creating within each chapter:

- In *The First Night*, the first mathematical concepts discussed are numbers in general. Robert raised the following question in response to the Number Devil informing him that there is an infinite number of numbers: "'How can you be so sure?' he asked. 'Have you ever tried?'" (p. 16). Similarly, Robert raised a question about the Number Devil's statement that there are infinitely many small numbers: "'The reverse? What do you mean?'" (p. 17). These questions asked by Robert may also be questions your students are considering. Before continuing to read the conversation between Robert and the Number Devil, take some time to allow students to think about these questions independently, then collaborate with some of their classmates on the perceived answers. Once students have had a chance to talk through the questions and their thoughts, challenge the students to connect these questions to real-world situations. For example, students might be able to connect this idea to perhaps a millionaire or billionaire who has more than enough money, but continues to make in a similar way that numbers continue. Finally, have students create a representation of the concept with their real-world connection to explain their findings.

- *The Second Night* continues to discuss numbers with a focus on Roman numerals before transitioning to a discussion of the number zero. "'Besides, what's so great about zero? Zero means nothing.'" (p. 36). This is a great place for students to collaborate, connect, and create. Ask students to collaborate in groups about what zero actually means and have them find real-world connections to the number zero. For instance, if you have zero dollars, you are not in debt nor do you have any money to spend. Or if there are zero minutes remaining for something to start, that means it is showtime. If there are zero copies of a library book remaining, then there are no copies to be checked-out. After students find connections, have them create a graphic or presentation for the class about their connection(s) they found.
- Also, in *The Second Night*, after discussing zero in great lengths, the Number Devil introduces the idea of hopping numbers. "'Hopping?' Robert said scornfully. 'What's that supposed to mean? Numbers don't hop'" (p. 38). Here students could be challenged to describe how numbers hop or what the Number Devil's hopping numbers might be as they collaborate with one another. Next, have students find where hopping numbers are in the real-world. Then have students create a task involving a real-world connection to hopping numbers for their classmates to complete. Hopping numbers represent exponential growth, whether it is the spread of a virus, sales of smartphones, or even population growth over given time periods. An extension to this exercise could include reading *One Grain of Rice: A Mathematical Folktale* by Demi (1997). This book demonstrates the quick affect that hoping numbers can exhibit if the king owes a servant one grain of rice on day one, then double the amount on each successive day for one month.
- *The Third Night* includes a discussion on division sparking the following: "'Because it's forbidden. Dividing by zero is strictly forbidden.' 'What if I did it anyway?'" (p. 54). Have students collaboratively try to divide by zero. Some students might be able to make the connection back to the inverse of division, which is multiplication, to help support the argument for not being able to divide by zero. Once they have reached a conclusion about dividing by zero transition to seeking a real-world connection. Here is an example of one such connection:
  - *If I have 10 Skittles, I can divide the 10 Skittles among my five friends. However, if I have 10 Skittles I cannot divide them among zero friends and myself. If there are zero people to share with then I have not shared my Skittles. Thus, it is impossible to divide by zero.*

Here students could illustrate their understanding and connections to share their findings with their classmates.

- In *The Fourth Night*, a lengthy discussion on decimals, approximation, and the infinite number of numbers ensues where the following two questions arise:

"'⅓ + ⅓ + ⅓ =1, doesn't it?'" (p. 71) and "'How many [numbers] would you say there are between 0.0 and 1.0?'" (p. 73). These questions invite thinking and conversation. Have students collaborate on what they think the answer to each of these questions might be. These questions might not be as inviting to real-world connections, but students could justify their thinking through visuals or words for a whole-class discussion. This is a place to let students create a "proof." It does not need to attend to the precision of formal mathematical proofs, but give students the opportunity to attempt to prove their case.
- *The Fifth Night* provides an opportunity for students to explore ideas on "quadrangle numbers" (p. 101) before the Number Devil reveals the answer providing the opportunity for students to collaborate and hypothesize what they imagine quadrangle numbers might be. In groups, have students draw what a quadrangle number would look like visually. Then continue reading as the Number Devil reveals the drawing on page 102.
- In *The Eighth Night*, there are multiple opportunities to interrupt the reading and ask students to determine the number of seating arrangements or handshakes for any number of students as the idea of combinations is explored. The dream focuses on those arrangements for two, three, or four people explicitly before extending beyond to other possibilities and introducing "vroom!," (p. 156), or factorial. In small groups, have students determine the number of handshakes or seating arrangements in a line that could be made with their group. Then challenge students to extend this to the whole class. After students have worked through these two tasks, ask students to find a real-world connection. Where does the number of combinations matter? Students may explore ideas of locks, license plates, or telephone numbers, just to name a few. Have students display their findings visually and then invite the class to partake in a gallery walk so students can explore the ideas of their classmates.
- In *The Ninth Night*, the Number Devil mentions series and Robert immediately asks, "'Series? What are series?'" (p. 180). Challenge the students to brainstorm ideas about what series are. After students spend five to ten minutes, allow students to utilize the internet to find other resources to verify their ideas and begin to understand the answer to this question. Then continue reading the chapter to see what Robert and the Number Devil have to say.

AFTER READING *THE NUMBER DEVIL: A MATHEMATICAL ADVENTURE*

The 13th Night Dream Sequence

Throughout the novel, Robert's dreams are focused on specific or sets of connected mathematical concepts. After reading the novel, either have the

Figure 9.5 Sample 13th Night Illustration and Caption. *Source*: Created by authors

students, individually, with partners, or with small groups create a 13th night. In their 13th night dream sequence, the students could focus on one or a small set of mathematical concepts just as the Number Devil did—a concept not covered in the novel is suggested. Students can get creative and create different terminology if they want, but they must be sure the concept is explained clearly through their writing, which should include dialogue as was modeled throughout the novel. Additionally, their 13th night can be accompanied by an illustration similar to those seen in the novel. An example of a 13th night illustration can be seen in figure 9.5.

After all students or groups have completed their 13th night assignment, display the illustrations throughout the classroom and have students go for a gallery walk to read and see what their classmates have come up with.

Making Mathematical Connections Visible

For this activity, groups of students could be assigned a different chapter from the novel (see figure 9.5). Each group should be tasked with creating a concept map of their night. The concept map should highlight the key ideas brought up in each night's dream so that the group can help take the lead in the whole-class discussion about their chapter (figure 9.6).

Certain chapters also have problems for the readers to try themselves (see *Activities Built into The Number Devil* section). These problems can be part of the group's responsibility to try on their own and then discuss together as a whole class. Specific examples of these problems are laid out in the next section. There may be some benefit to doing an example concept map together as a class for one of the first chapters so the students have an idea of what they should include as they create their own concept maps.

Dream	Concept	Brief Explanation
The First Night (pp. 7–26)	Infinity	There is an infinite amount of numbers. You can add one to itself forever. There is no end, and you cannot count to it.

Just as there are an infinite amount of big numbers, there is an infinite amount of small numbers. You can split one object forever into smaller and smaller fractions. |
The Second Night (pp. 27–46)	Roman Numerals	I is used for 1, X is used for 10, L is used for 50, C is used for 100, and M is used for 1,000. If a letter comes before a number, you subtract it (CM=900). There was no zero
	Zero	Zero was the last number to be discovered
	Raising to a Higher Power	When you multiply a number by itself a certain amount of times, x, you raise it to a power of that number, x (5x5x5=5³)
The Third Night (pp. 47–64)	Division	Dividing is the opposite of multiplication. Unlike multiplication, addition, and subtraction, you will not always get a whole number
	Dividing by Zero	Cannot divide by zero; you cannot break up a number into zero parts
	Prime Numbers	Numbers that can only be divided by 1 and itself
The Fourth Night (pp. 65–86)	Place Value	The value of a digit in a number based on its location in that number, the US uses commas and decimals to help identify place value, the Number Devil uses commas instead of decimals when he writes out numbers as this is how they are written in Europe
	Repeating Decimals	A decimal that has the same set of digits that repeat indefinitely, this could be one digit that repeats or a set of digits that repeat, for example 1.66666 . . . or 1.134513451345 . . .
	Irrational Numbers	A number that does not play by the rules; a number that cannot be written as a fraction
	Square Roots	Square root is the inverse of squaring, taking the square root of a number means finding a factor of the number so that when that factor is multiplied by itself you get the original number
The Fifth Night (pp. 87–104)	Triangle Numbers	A series where you continuously add the natural numbers as illustrated by constructing triangles where each row is one unit greater than the next (1, 3, 6, 10, 15, 21, . . .).
	Quadrangle Numbers	A series where you continuously square the natural numbers as illustrated by constructing squares where each side length is one larger than the previous (1, 4, 9, 16, 25, 36, . . .).
The Sixth Night (pp. 105–122)	Fibonacci Numbers	A sequence of numbers where the next term is the sum of the previous two terms (i.e. the Fibonacci numbers for n=1 would be 1, 1, 2, 3, 5, 8, 13, 21, 34, . . .)
The Seventh Night (pp. 123–146)	Pascal's Triangle	A triangular array created through the summation of adjacent terms of the previous row

Dream	Concept	Brief Explanation
The Eighth Night (pp. 147–168)	Factorial	A mathematical expression marked by an exclamation point that means to multiply the given number by each digit between 1 and that number. For example, 4! (4 factorial) means 4*3*2*1
	Combinations	The number of possible arrangements of objects where order does not matter
	Permutations	The number of possible arrangements of objects where order does matter
The Ninth Night (pp. 169–188)	Series	A series is the sum of the numbers or values in a sequence
The Tenth Night (pp. 189–210)	Geometric Shapes	An enclosed area created by a varying amount of sides, angles, and side lengths
The Eleventh Night (pp. 211–230)	Raise to the Zero Power	Every number raised to an exponent of 0 gives a result of 1
	Proofs	A proof is a mathematical justification or argument for why something works the way that it does. Proofs rely on making a series of logical statements to get from some beginning assumption to a guaranteed conclusion
The Twelfth Night (pp. 231–253)	Mathematicians	A person who studies math. This could even be you!

Figure 9.6 Mathematical Concepts Explained. *Note:* This figure includes key concepts found within Robert's dreams. The concept is listed under the first dream it appears in, but some concepts are revisited in subsequent dreams

Bringing the Mathematical Conversations Full Circle

In order to gauge students' growth of knowledge of key mathematical ideas after reading *The Number Devil*, the same baseline knowledge pretest can be given as a post-test. Teachers and students will be able to see where the class gained or clarified their understanding of specific concepts at the end of the unit. Sharing the data with the students will also be helpful, as they realize that reading the novel did indeed contribute to their mathematical understanding.

Similar to the baseline assessment and the vocabulary knowledge-rating chart, chalktalks can be utilized both before and after reading *The Number Devil*, in addition to while reading. Chalktalks provide an outlet for students to demonstrate their knowledge or clarify their understanding by posing questions or processing other classmates' questions. After reading the novel, the same questions can be used or maybe a new question will come to mind while reading *The Number Devil* based on the particular students with the class. Here are some higher order prompts:

- Explain the different classifications of numbers.
- Where are geometric shapes in the real-world? Provide some examples.
- Where/How do you use mathematics in the real-world? Be specific with examples including where and how the mathematics is used.

If the initial chalktalks were completed on poster paper, or if they happen to still be on a board in the classroom, students could revisit that same question and add or clarify what was previously written. Including a chalktalk at the end of the unit can bring closure to the novel study as well as allow students to see their growth, if any. Clearly, any strategy used before reading and instruction began can be returned to as a way to see student growth over time.

BEYOND *THE NUMBER DEVIL:* A MATHEMATICAL ADVENTURE

Meet the Mathematicians

During Night 12, Robert is invited to Number Hell/Number Heaven where he encounters the mathematicians Lord Rustle (Russell Bertrand) (pp. 224–226, 238–239), Dr. Happy Little (Felix Klein) (pp. 239–240), Professor Singer (Georg Cantor) (pp. 240–241), Owl (Leonhard Euler) (pp. 241–242, 244), Professor Horrors (Carl Friedrich Gauss) (pp. 241–242, 244), Bonacci (Leonardo de Pisa) (pp. 108, 242, 244), Pythagoras (pp. 242–243), and Archimedes (p. 246). Also mentioned throughout the book are Blaise Pascal (pp. 128–145) and Man in the Moon (Johan van de Lune) (pp. 222–223). While a little bit about each of these mathematicians is mentioned, there is still much to learn. Allow students to pick one of these mathematicians and research them further in order to learn more about their work and lives. Students can work on their own or in groups in order to present more information on their mathematician. Some of the ways this can be done is through writing a children's picture book about them, filming a video documentary, or creating an interactive PowerPoint presentation.

Gender and Math

Additionally, in the story, Robert notices that all of the mathematicians are men (p. 245). Encourage students to research mathematicians who did not appear in the story, focusing on women or marginalized people. They should look through a few mathematicians to find ones with whom they can personally connect. Students may honor women mathematicians such as Hypatia, one of the first ones known; Sofya Kovaleevskaya, the first woman to obtain a doctorate; Emmy Noether or Ada Lovelace. Students may relate to black mathematicians such as Kathrine Johnson, who worked for NASA; Benjamin Nanneker, a freed slave with smarts in many other subjects; or Marjorie Lee Browne, who was a math teacher, Finally, they may want to look into eccentric mathematicians such as Paul Erdös, a man with a brilliant mind who hat to rely on the mathematicians he worked with to take care of him, or Kurt Gödel, who died tragically after starving to death after being convinced people were trying to poison him.

Students may present on these mathematicians in the ways mentioned before, or they may write them into *The Number Devil* by writing a segment for Night 12, including a fictitious name, description, and activity for them to be doing. Compile the segments together to create a new Number Hell/ Number Heaven meeting created by your class.

Mathematically Minded Book Clubs

Another possible extension activity for students is the incorporation of other novels into the math curriculum through book clubs with mathematics themes. Students could be given text choices exploring similar curricular concepts and self-select books to explore with classmates. Once students have had an experience with seeing mathematics in the literary world, they may find they enjoy learning about real mathematical concepts in the not-so-real world. There are many great choices for novels, many of which are explored in the other chapters of this book. Here are just a couple more suggestions:

The Boy Who Loved Math: The Improbable Life of Paul Erdös by Deborah Heiligman. This storybook describes the life of the prolific mathematician Paul Erdös. Students of all grades will be inspired by his lifestyle, which proves anyone can be a mathematician, and be invested in his adventures of jumping from country to country working on math with hundreds of great minds from all over the world. More advanced math students can investigate the illustrations in order to learn more about the mathematical concepts that he worked on such as the party problem and proof of Chebychev's theorem.

From telling us the remarkable story of his childhood and describing his unusual adulthood of whimsy and dependence on others, to incorporating Erdös's math through short quips and illustrations, this book proves that a picture is worth a thousand words.

The Curious Incident of the Dog in the Night Time by Mark Haddon. This novel follows the adventure of a young man, Christopher, who sets out to investigate the death of his neighbor's dog. While learning how to navigate the train system, work on expressing his emotions, and getting used to being on his own, Christopher integrates plenty of math concepts into his narrative. In fact, Christopher tries to see the world through a mathematical lens, because he feels that it is more logical to understand that way. Students will be exposed to prime numbers, the Monty Hall problem, solving for triangle side lengths, and more while connecting with this likable character on his journey of math and personal development.

CONCLUSION

The Number Devil is a novel that can be incorporated into any mathematics class, regardless of the level. This novel can act as an engaging entry point for students who lack confidence in their mathematical abilities or who are still working on developing their mathematical skills. The Number Devil introduces mathematical concepts through storytelling and invites the reader on an exciting adventure creating a sense of playfulness not always associated with the hard work of problem-solving in the mathematics classroom. The conversational, creative language serves to make a variety of mathematical concepts more accessible for students who claim that math isn't their favorite subject. The initial concept accessibility, in turn, helps students take a relatively basic level of mathematical understanding and delve deeper into the concepts. Teachers can use the before-, while-, and after-reading activities presented here as a general guide for using *The Number Devil* in the classroom with confidence that each of these strategies will help build mathematical literacy and academic language.

REFERENCES

Anthony, H. G., Kolodziej, N., & Meadows, J. R. (2017). From disenchanted to intrigued: Unveiling the *Number Devil's* tricks in precalculus and calculus. In P. Greathouse, J. F. Kaywell, & B. Eisenbach (Eds.), *Adolescent literature as a complement to the content areas: Science and math* (pp. 147–161). Rowman & Littlefield.

Demi. (1997). *One grain of rice: A mathematical folktale*. Scholastic Press.
Enzensberger, H. M. (1997). *The number devil: A mathematical adventure*. Picador.
Kilpatrick, J., Swafford, J., & Findell, B. (2001). *Adding it up: Helping children learn mathematics*. Research report for the Mathematical Learning Study Committee, National Research Council. The National Academies Press.
National Council of Teachers of Mathematics. (2000). *Principles and standards for school mathematics*. https://www.nctm.org/Standards-and-Positions/Principles-and-Standards/.
[tennisdog2468]. (2008, May 5). *Number devil* [Video]. YouTube. https://www.youtube.com/watch?v=qJHc54IG5R8.
Vygotsky, L. (1978). *Mind in society*. Harvard University Press.

Chapter 10

Matchmaking Mathematics
Teaching Algorithms and Probability with Nandini Bajpai's *A Match Made in Mehendi*

Jen McConnel and Allen Harbaugh

Although algorithms do not always require numeric concepts as part of their programming, the dating apps that have proliferated in recent years offer diverse combinations of variables and percentages in order to help individuals increase their probability of making a match. However, algorithms are not free from bias, and it's hard to really typify the algorithms driving these apps as neutral or objective. But that doesn't stop both real people and fictional characters from putting an intense amount of trust in these algorithms, including the characters in Nandini Bajpai's (2019) contemporary young adult novel *A Match Made in Mehendi*.

Protagonist Simi teams up with her tech-savvy brother and her popularity-focused best friend to convert her Desi family's ancestral matchmaking wisdom into a matchmaking app targeting teenagers. The team decides to pilot the app at their own high school, with mostly positive results. Simi puts all her faith in the algorithm, so much so that she is willing to ignore her own attraction to someone the app does not match her with. The novel has a satisfying conclusion that affirms the value of the app and its algorithm, but also emphasizes the human variables that no algorithm can plan for—proximity, chemistry, and individual growth.

Bringing this novel into a secondary mathematics classroom opens up a range of possibilities. This chapter emphasizes the ways in which Bajpai's novel can be integrated into conversations about algorithms, probability, and variables by focusing on the following driving questions: *Which variables affect authentic relationships? And, how can you measure and plan for these variables?*

A MATCH MADE IN MEHENDI BY NANDINI BAJPAI

Simi Sangha, 16 years old, is ready for her sophomore year to be different from previous years. With her best friend, Noah, she sets out to build her confidence and express her true self. But while Noah dreams of a year of popularity, Simi just wants to make art, avoid the mean girls, and figure out how to tell her mom and aunt that she is not ready to be the next in the long line of family matchmakers. But Noah thinks the matchmaking tools are brilliant and a sure path to high school stardom for them. Noah and Simi enlist Simi's older brother and engineering whiz Navdeep to help them launch an app based on the family business. Navdeep has already programmed a basic algorithm based on the matchmaking guide their mother cherishes, but he was not able to pilot the app because his mother and Simi's mother did not want to lose the interpersonal element of the matchmaking business. With Noah pushing for popularity and Navdeep eager for a chance to actually test his creation, Simi begins to get excited, and together they launch "Matched!," a savvy dating app that is limited to students at their high school.

At first, the app is a success. Students are matching and forming connections across clique boundaries, and Simi and Noah find themselves in the center of a growing group of kind, funny, and supportive friends—students they knew before as acquaintances but never took the time to get to know deeply. Simi puts her faith in the app, trusting her brother's programming and her mother's matchmaking secrets over her own attraction to Suraj, a new boy at school. Ultimately, though, Simi decides that although she trusts the app, she also trusts her own instincts and she begins to consider that she could enjoy being the next *vichola* in her family.

BEFORE READING *A MATCH MADE IN MEHENDI*

As an entry activity to this book teachers could ask students to work collaboratively to develop a list of variables that affect relationships. As a whole class, come together virtually using Padlet, or in person using chart paper. Invite students to add different kinds of relationships to the Padlet or chart paper as a first step. If prompting is needed, the teacher could reference different types of relationships, such as parent and child, friend and acquaintance, teacher and student, romantic partners, and employer and employee. Once a list is generated, students can be split into groups to take a deep dive into the variables of one of the specific relationships listed. For example, the variables they might brainstorm for the relationship between parent and child could include age, people in the household, presence or absence of other children, presence or absence of family members, culture, values, and beliefs.

After groups have brainstormed categories, guide them on a virtual carousel walk, visiting other relationships on the Padlet and adding other variables that they think of. Finally, the class will come back together for a summative whole-class discussion. The teacher could facilitate either a conversation or a reflective writing prompt by asking questions to help students make connections between the variables they identified:

- What similarities do you notice between these variables?
- What patterns are you noticing here?
- In what ways do some relationships have more variables than others?
- If you were asked to explain this list to someone outside this classroom, how would you explain it?
- Based on this work, what variables do you think are most important for any type of relationship?
- In what ways does the type of relationship change the variables you would consider important?

Interpreting Probabilities

Before reading, a discussion about the nuances of probabilistic reasoning—the idea of thinking either in extremes (for a sense of certainty) or counterfactually (to gauge uncertainty)—sets the stage for students to more critically assess how the idea of percent-match and probability have been conflated in the novel. As a way to approach this, we have suggested a three-part plan. The first part engages students in thinking about how rare an event needs to be for its occurrence to be considered unusual. In the second part, students begin to quantify their own metric for unusual versus usual. The third part is to mathematically explore (with the assistance of a random number generator) the range of probabilities necessary for an event over multiple observations to be unobserved.

Part I: When Something Is Unusual (Chapter 2)

The objective is for students to reflect on what they consider unusual, and to recognize that their cutoff for classifying something as unusual will be different from other people's cutoff values, and that such cutoffs are usually arbitrary.

Thought Experiment: Twins? Have students consider the following scenario:

> You are buying a meal at a fast-food restaurant, and you chat briefly with the cashier. They mention that they go to a school near you. You know some people

at that school, but you've never met this person. For conversational purposes, you ask if they have an older sibling that may have known a friend of yours. They tell you no, and they mention their only sibling is their twin.

Ask students to consider the frequency with which multiple births occur. Roughly 1 in 90 births is a multiple birth, and 1 in 250 births is identical twins (*Twins Magazine*, n.d.). Triplets occur 1 in 4,000 births, and identical triplets are estimated to occur 1 in 1 million births (Geggel, 2013).

Part 2: Exploring Your Cutoff for Unusual
(vs. Not Unusual) (Chapter 6)

The teacher can introduce this discussion by using a script such as the following:

> Even if you don't know it, you have some metric you use to decide if something is unusual or usual. First, let's explore the idea that you do indeed have some boundary where you transition from thinking something is unusual to usual.
>
> Let's say you think it is unusual to meet a stranger and learn that they have an identical twin. One stranger . . . that's unusual. But, if you were at a sporting event with 30,000 fans in the stadium, you probably would not be surprised to learn that there is at least one person in the crowd who has an identical twin (they may even both be there for the game). If you are surprised when it is only 1 person, you might not be surprised if it is 30,000 people. So, what is the number where you switch? In student groups, discuss a range of numbers: 2 people, 5 people, 10 people, 25 people, 100 people, 250 people. Can you locate your number? Can you locate a number that works for your entire group? Context is also important to consider here: what you are thinking about when you decide if something is unusual is influenced by your context at the time.

Here is another example to prompt discussion about the ways in which people arrive at their cutoff points for labeling something as unusual. Ask students to think about a trip they have taken on a fairly regular basis (ideally daily or so). For example, prompt them to think about their walk or drive to school from home. Let's say they drive (or are driven). Ask them to consider the following question: *How many traffic lights do you encounter along the trip?* Once a number is determined, have students consider how many of these lights they would expect to be red/stop on any one trip. Encourage them to think about the consistency of this: *Is it always about this number? Does it seem to vary quite a bit? Either way, why?* Now ask students to imagine that on their trip tomorrow they encounter all red/stop lights: *Would this be* unusual? *(Or said another way, would you be surprised if this happened or not?).* Now position students in the opposite direction, headed home from school. Pose the following question: *Would you be surprised if you encountered absolutely no red/stop lights?* Now the fun part, have students think through the range of usual lights they encounter that are red/stop. Said another way, what is the cutoff for too few (or too many) red/stop lights to be considered usual or unusual?

Part 3: Reasonable Probabilities for Unusual Events (Chapter 7)

The tricky part about thinking probabilistically is that you are using probability to predict a single event, but probabilities are focused on the outcomes of many events, or the same event being repeated many times. Let's say the chance of winning a video game is 10%. If there are 20 students in class, and each student plays the game, would it be unusual for someone to win? 10% is a pretty low number, but if we let 20 students play the game, we could end up with the following results:

0 0 0 1 0 0 0 0 0 0 1 0 0 0 0 0 1 0 0 0

A one means the student won, and a zero means they lost. Here, by random chance, 3 of 20 students won. (A simulation such as this can be obtained by typing = (RAND () < 10÷)*1 into any cell in Excel and then copying it into 19 other cells.) In fact, it turns out that the probability of none of the 20 students winning is 12.16%. Stated another way, the probability that at least one of the 20 students win is actually 87.84%. So, even though the probability of success is only 10%, in a group of 20 students, this means it will happen at least once.

Now, let's say the probability of winning is 5% (obtained in Excel by typing = (RAND () < 5%)*1 twenty times):

0 1 0 0 0 0 0 0 0 0 0 0 0 0 0 0 0 0 0 0

Still, someone won. What about only 1% (obtained in Excel by typing = (RAND () < 1%)*1 twenty times)?

0 0 0 0 0 0 0 0 0 0 0 0 0 0 0 0 0 0 0 0

At 1%, no one in 20 students won the game. The probability of no wins in 20 students is 81.79%.

Probabilistic thinking requires thinking about counterfactuals, the idea of "what-if" this happened again and again. And, the idea of certainty that something will (or will not) happen doesn't just require a small probability, but a very small probability, particularly if you are going to be doing the activity repeatedly.

WHILE READING *A MATCH MADE IN MEHENDI*

Although the central story of the novel focuses on the matchmaking app that Simi, Noah, and Navdeep launch at their high school, the book is full of relationships that move beyond romantic matchmaking. Simi's relationship with her family is rich and explicitly present throughout the text, including the nuances in her sometimes-strained relationship with her mother compared to the deep connection she feels with her grandmother, as well as her relationship with her cousins and extended family. Then there are the relationships Simi builds at school over the course of the novel: her best friendship with Noah, of course, but also her growing friendships that come about as a result of Matched! Teachers can emphasize different relationships through the reading, connecting back to the lists students brainstormed before starting the novel, to guide students through an exploration of the complexity of relationships presented in the text. Some questions teachers might ask during reading include the following: *What is the most important relationship Simi is involved in? Why do you think it is important? How is the Matched! app facilitating relationships beyond romantic pairings? How does Simi act and participate in her various relationships? Is she consistent throughout her interactions, or does she shift her behavior based on the context of the relationship? What does Simi's behavior in various relationships tell us about her character?*

While reading the novel, students could also keep track of the variables that Simi includes in the Matched! app. The excerpts from the Shagun Matchmaking Guide that start some of the chapters will provide some of the variables, and a close reading of the text will illustrate others. For example, a student might make note of the following variables: sense of humor (p. 69), favorite food (p. 74), bravery (p. 75), honesty (p. 75), and environmental consciousness (p. 76).

As students encounter the variables Simi places at the heart of the matchmaking app, they can compare these to the variables they generated as a class in the before-reading activity. The list can be maintained in a virtual collaborative space, like a Google Doc, or on a piece of chart paper in the classroom. Once there is a list of at least five variables to compare to the students' lists, the teacher might ask the following discussion questions: *What do you notice about your list of variables in relation to Simi's list? What surprises you when looking at these two lists? What revisions to your variable list would you like to make?*

Predictions for a Successful Pairing

The objective of this activity is to expose students to the process of modeling random events. This is the idea that the probability of a random event (the success or failure of a pairing of two individuals on an app) is a value that can be obtained with a mathematical calculation.

As with the before-reading activity about interpreting probabilities, this activity is broken into three parts. The first part is for students to use a match-score to predict the probability of a successful match using a logistic curve function. The second part is to examine how small changes in a match-score may or may not result in a substantive change in the predicted probability of a successful match. The third part is to connect the ideas from the first lesson to better understand the limitations of an increase in a match based on where the match-score is located.

Part I: Converting a Match-Score to a Probability of Successful Pairing (Chapter 11)

This part of the activity introduces the logistic function. The degree to which the concept of function and graph has been utilized will impact the level of exploration possible. What is presented here is based on the assumption that functions may not be fully covered, but the idea of graphing bivariate relationships and reading such graphs has been covered.

The Logistic Function. The logistic function is a tool that is used quite frequently in mathematics and probability. Because probabilities can range from 0% (something that is impossible) to 100% (something that is certain to happen), it is important to have a function or procedure that only produces probabilities in this range, that is to say, between 0% and 100%.

One type of logistic function can be given by the formula where m is the number of matches on a 20-question quiz (like the one in the app from Chapter 7 in the novel):

$$p(m) = \frac{1}{1 + 2^{15-m}}$$

We will call m the match-score. As the number of matches increases between two individuals, we would expect the probability of a successful pairing to increase. And this can be seen in the graph for this function in figure 10.1.

Reading the Graph. From the graph, we can use the number of matches between two people's survey/quiz to predict the probability that their pairing would be successful. Let's say two people only matched on five of the questions. If we plug this into the function, we would get $p(m) = 1/1 + 2^{15-5} = 1/1 + 2^{10} = 0.000976$. This is a very small probability

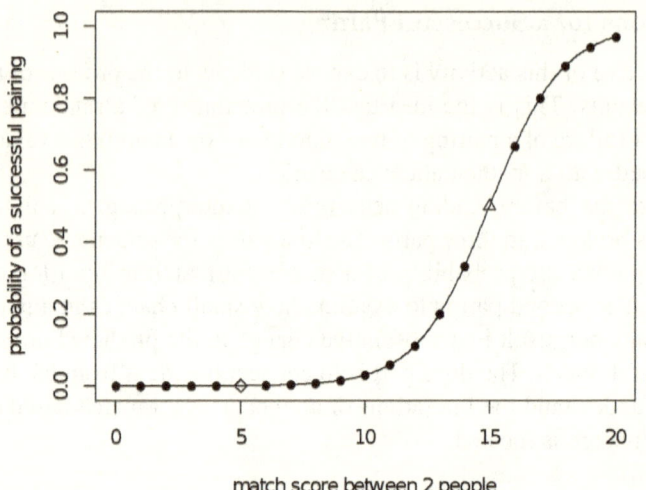

Figure 10.1 Relationship between Probability of a Successful Match and the Pairs Match-Score on the Quiz. *Source*: Created by authors

of success, in particular it would be a successful pairing only for about 1 in 1000 pairs. From the graph, we can see this as the diamond point on the curve. We notice that the predicted probability is very close to zero. Thus, it is very unlikely this would be a successful pairing.

But, what if two people matched on 15 items: $p(m) = 1/1 + 2^{15-15} = 1/1 + 2^0 = 0.5$. In this case, there is about a fifty-fifty chance that the pairing would be successful as visualized in the graph by the triangle point on the curve. Recall from the first lesson that this does not mean the match will happen, but it does suggest that it would be particularly surprising if it did happen, unlike when there were only 5 correct matches, in that case, a successful pairing would be rather surprising.

The "Algorithm" for This Logistic Function. The way that this function/procedure turns a match-score into a probability can be broken down into a step-by step procedure. Let's use the last calculation as an example:

1. Subtract the number of matches from 15: (15 − 15 = 0)
2. Raise 2 to the result from Part I: (2^0 = 1)
3. Add 1 to the result from Part II: (1 + 1 = 2)
4. Take the reciprocal of the result from Part III: (½) = 0.5

Next, have students calculate the predicted probability of a match for different values of match-scores. Have them interpret the probability and reflect on whether it would be usual or unusual if two people with their match values successfully paired together or not.

Part II: How Changes in Match-Scores Result in Changes in Probabilities (Chapter 22)

This part of the lesson shows students that some range of values become something of a "dead-zone" for successful pairings. This helps us understand why Amanda couldn't change her responses to improve the present pairing with Ethan.

Some Changes Don't Have as Much as an Impact as Others. First, explore what happens if two people have a weak pairing. Let's say two people have a match-score of only three. The probability of a successful pairing is 0.02%. This is a very low success probability. But, what if they had a better match-score, say they increased to six. Well, unfortunately, that only results in a success probability of 0.2%. Still, a very unlikely chance for a successful match. Ok, but what if they were able to change their answers so they matched on 9 of the 20 items. Well, it still only results in a 1.5% chance of a successful pairing. This means only 3 out of 200 such matches would ever result in a successful pairing.

It All Depends on Where You Start. So, we just saw that there wasn't much change if we started with a really low score. But what happens if we start with a slightly better score? Let's say we start with a 10 and we increase by 9 points. In this case, we go from a 3.0% chance of a successful pairing to a 94.1% chance of a successful pairing.

One interesting avenue for discussion is to think about how a "close" match could still be a match. Let's use an overly simplified example. Ask students to rate their agreement with the following statement: Pistachio flavored ice cream is the best flavored ice cream. *Do you agree? How strongly do you agree? (rate as 7, 6, or 5). Do you disagree? How strongly do you disagree? (rate as 1, 2, or 3). Are you neutral, without an opinion either way? (rate of 4).* It is a good discussion topic to consider how something rated as two on this scale can be interpreted as a match in one setting, but not in others. For example, one might argue that (slightly agree) and (strongly agree) are a match because they both agree. But it would be nearly impossible to say that (slightly disagree) and (slightly agree) are a match in any sense (but they are both only two "spaces" apart on the Likert scale).

After explaining the above information, teachers might ask students to look at how two, four, or six points change the overall outcome for different starting points. Students could also start with a high match, and show how decreases in the match-score can have a much stronger impact. To consolidate learning from this activity, the teacher could ask the students to interpret the probabilities using the probabilistic reasoning they worked to establish in the first lesson.

Part III: Not Much Change for Amanda (Chapter 24)

This part of the lesson asks the students to mathematically explain why Amanda was unsuccessful in changing her prediction probability with Ethan when she kept trying to change her answers—even if she did stumble upon a match, their starting position for the match-score was so low, it had little-to-no impact on changing their rating on the app.

Wrapping Up. The key for this lesson was to think about how the match-score was converted into a probability for a successful pairing. However, the idea of a matching app is far more complicated than just asking how many questions people answered the same. Have the students reflect on the different ways a "match" could count as a match. This is a good place to begin to critique the book with regards to "matching" with open-ended questions such as the following: *If Ethan and Téa had written their list of favorite soccer players in a different order, how do you think that would have affected their match?* (p. 149). *Why did characters like Aiden and Amanda lie on their answers?* (Chapter 21). *What might some of the variables be that give Simi a strong match with Suraj?* (p. 310).

AFTER READING *A MATCH MADE IN MEHENDI*

By the end of the novel, Simi's perspective on matchmaking has changed. She no longer looks askance at the family tradition, but like her grandmother, she views matchmaking as a form of community building. Not only has she found that she enjoys helping bring romantic couples together, but she also recognizes that the process of working through the app has shifted her relationships, bringing her together with a group of friends across cliques who support each other in crisis and add a lot of laughter to her life. And then there are the relationships with her family, strengthened through her understanding of matchmaking and acceptance of its role in her own life.

Students can return to and answer the driving questions at the heart of this chapter—*Which variables affect authentic relationships? And, how can you measure and plan for these variables?*—through a literary analysis response that focuses on one relationship of their choosing from the text. Bring students back to the conversation they participated in before reading where they brainstormed the variables that they found important in different relationships. Ask the class to create a list of relationships in the novel, and classify the type of relationship. For example, such a list might include the following:

- Simi and Aiden (a friendship and a match gone bad)
- Amanda and Simi (enemies)

- Simi and Noah (best friends)
- Simi and Navdeep (siblings)
- Simi and her grandmother (family)
- Simi and Suraj (friends to something more)
- Ethan and Téa (a strong match)
- Simi and Jassi (from acquaintances to friends)

Using the relationship of their choice, students could respond to the driving questions of this unit by developing a multimedia composition that draws on the content of the novel to determine which variables affect authentic relationships and ways that these variables could be planned for. For example, Simi and Jassi share a faith and out of school cultural group as members of the Sikh and Desi community, but until the matchmaking app, they had not ever connected beyond being polite to each other. However, Jassi shares Simi's love of henna and is the first person to ask Simi to do henna for her. Their values about honesty and friendship align, with Jassi perhaps even feeling more strongly than Simi, since Jassi continues to reach out and support Simi, who initially struggles with asking for help from the people like Jassi that she eventually considers friends. And Jassi matches with Suraj, although not as strongly as Simi does, indicating that there is even more overlap between their personalities than is explicit in the text. There is a lot to unpack in their relationship, and in any of the character relationships in this novel.

Student compositions can take many forms, depending on the context of the classroom: students might make a slide presentation or a video response, or students could experiment with an online story platform such as a choose-your-own narrative composer like Inklewriter. Regardless of the form students choose for their composition, they should include the following elements:

- An explanation of the relationship they have selected (who is involved and what kind of relationship it is);
- Textual evidence that specifically addresses the driving questions;
- Analysis and synthesis that demonstrates the students' own thinking about the driving questions in relation to the text.

Algorithms: How Matches (or Mismatches) Become Probabilities for Successful Pairings

The idea of this after-reading activity is that it shows that the algorithm used to make the probability predictions are far more complicated than can be easily expressed in a novel or with a simple quiz. Simi, Noah, and Navdeep put a lot of thought into translating the values-based scenarios in the family matchmaking guide into quiz questions (Chapter 7), but although the novel explores the questions they ask, it does not delve into the programming behind that app that actually makes the matches.

The first thing to emphasize to students is that the match-score, as we worked with during reading, is much more complicated. Previously, students were asked to consider the number of items on a survey the two people agreed upon. In truth, it is more complicated than this because there are some questions where you might want people to agree, but there are others where you want them to disagree, the idea of "opposites attract" (in a mathematical sense).

Also, there are some items that probably carry more weight than others. For example, a question about favorite ice cream may not be as important as a question asking about whether you support protests during the national anthem at sports games. So, what happens is that we don't use just the number of questions that match, but we use a more complicated formula to turn the responses from each pair of people into a match-score. It is this match-score that is plugged into the logistic function, and this is what gives the predicted probability for a successful pairing.

Measuring Variables

Teachers can ask students to revisit their list of variables from earlier in the unit, and to work in pairs to create questions to measure the different variables. Once students have written sample questions, the class can come together to discuss whether the match-score should be higher or lower based on agreement or disagreement. Finally, teachers can ask students to rank the items based on which variables they think should have more of an impact on the match-score, working through a process similar to the one Simi, Noah, and Navdeep undertook to convert the family matchmaking guidelines into a high school appropriate app in Chapter 7.

BEYOND *A MATCH MADE IN MEHENDI*

Simi's interest in the Matched! app is less focused on the Science, Technology, Engineering, and Mathematics (STEM) elements of algorithms and app design, and more focused on the human variables at play. But even before that, her first moment of excitement with the idea comes from the opportunity to use her art to create icons for the Matched! Profiles (pp. 75–76). However, as Simi begins to learn, creativity is not absent from the STEM fields her brother and Suraj enjoy. This subtle shift in the narrative, from STEM to STEAM (adding Arts to the metaphorical equation), offers an opportunity for extension activities that invite students to consider the relationship between mathematics, science, and art.

Algorithms beyond the Novel

If it's slightly more nuanced than the book can really explore, how are the matching-scores calculated in the real world? Answering this question requires the collection of a lot of preliminary data. This data has the responses on the survey for each pair, but it also has an indicator for if the pairing was successful or not. This data is used to build a function that turns the responses into a match-score, and then turns the match-score into a probability for a successful pairing.

The app is then released to the world. However, data collection does not stop after an app is released. As consumers use the app, the app developers update the algorithm. Or, more likely, a computer is programmed to make updates to the algorithm. This is an example of matching learning, or what some call AI, artificial intelligence. As more and more people use the app and their results are tracked over time, the hope is that the updates to the algorithm get better and better. Presumably, Navdeep will continue to refine the algorithm after the pilot at their high school, in order to better maintain a tool that will be useful in the family matchmaking business, but it is good to remind students of the long-term aspects of designing and running a successful algorithm that are not addressed within the novel.

Modernizing a Tradition

For this extension activity, invite students to update something from a family or community tradition and modernize it in some way. Simi took the big book of matchmaking and made an app. In similar form, students might choose to revisit their grandmother's recipe for turkey stuffing and make it vegan, or consider a local celebration that could integrate digital components (What about a Zoom parade?). Students can present their update to the class and discuss the process that went into updating the tradition. For a written component, have students write an artist's statement that explains the choices they made.

Life Imitates Art

In an exploration of the way life imitates art, students could listen to the short NPR story titled "Senior at N.H. High School Wins Top Prize in Science Talent Search" (2021) and write a fictional exchange between real-life app designer Yunseo Choi and Simi. *What might they say to each other or ask each other? What might Simi learn from Yunseo's prizewinning experience?*

Where Else Can Algorithms Be Used?

Another excellent extension activity is one that makes space for students to explore other places algorithms like this might be used, such as predictive policing or loan approval processes (as introduced in Cathy O'Neil, 2016). This also could be a springboard for a discussion on how there could be deeply flawed algorithms at play in the world right now, perhaps augmented by sharing an excerpt from *Weapons of Math Destruction* by Cathy O'Neil (2016). This text, which discusses the darker side of algorithms as they have been applied in situations of predictive policing and loan approval, is an accessible read for upper-year students who want to take a deeper dive into understanding algorithms and their real-world applications following this unit.

CONCLUSION

Using *A Match Made in Mehendi* in the context of a secondary mathematics classroom makes space for conversations about algorithms, probability, and variables that expand beyond the course, helping students make real-world connections to the content of their math lessons and their own lives as they work to answer these driving questions: *Which variables affect authentic relationships? And, how can you measure and plan for these variables?* Like her grandmother, Simi eventually comes to the realization that matchmaking is about more than tradition: it is about strengthening the relationships that hold a community together. The relationships in Bajpai's novel are rich and nuanced, and even though the book focuses on matchmaking, it is more than a standard rom-com. It is a STEAM exploration of the ways strong relationships can be formed, and it offers a great deal of possibilities for discussion and higher-level thinking when used as a course text.

REFERENCES

Bajpai, N. (2019). *A match made in Mehendi*. Little, Brown and Company.
Geggel, L. (2013, May 26). *Oh baby! The science of identical triplets and quadruplets.* Live Science. https://www.livescience.com/52613-identical-triplets-quadruplets-science.html.
O'Neil, C. (2016). *Weapons of math destruction.* Crown Books.

NPR. (2021, March 18). *Senior at N.H. high school wins top prize in science talent search*. Morning Addition. https://www.npr.org/2021/03/18/978495989/high-school-senior-wins-top-prize-in-science-talent-search.

Twins Magazine. (n.d.). *Incidence of twins by twin type*. https://twinsmagazine.com/incidence-of-twins-by-twin-type/).

Chapter 11

Simulating Success Based on Societal Odds

A Mathematical Read of The Hunger Games

Robin Keturah Anderson, Melissa Troudt, Candace Joswick, and Lisa Skultety

Societal pressures influence almost all aspects of a young adult's life. From their day-to-day choices, such as fashion, sports, or friends, to their long-term career aspirations, young adults are constantly assessing how they will be perceived by society. Whether it is following the fashion trends of the most recent social media influencer, or pursuing a career in computer science because they can likely make a lot of money, young adults make choices that will impact the way that society views them. This chapter interrogates some fundamental questions around choices in society. *What advantages are afforded to someone who conforms to societal expectations? What freedom does one have to influence their chances for desired outcomes in society?* Through activities tied to Suzanne Collins's young adult novel *The Hunger Games* (2008), students will use mathematics to investigate how societal expectations manipulate citizens' opportunities for success, which inevitably create and perpetuate inequality in society.

"May The Odds Be Ever In Your Favor."

Effie, District 12 Escort, *The Hunger Games*

THE HUNGER GAMES BY SUZANNE COLLINS

In *The Hunger Games* two children from each of the country's (Panem) twelve districts are chosen as tributes for a to-the-death survival competition, called the Hunger Games. The book centers around Katniss, a 16-year-old young woman from District 12, who volunteers as tribute to replace her younger sister. The Games are televised to the Capitol and all districts; Gamemakers determine the terrain of the Games and rules—which can change during the Games, and tribute sponsors procure food and tools delivered to competitors throughout the Games. The book begins with the reaping or selection of tributes (Chapters 1 and 2), follows the training and preparation of tributes for the Games (Chapters 5–10), and chronicles the events that lead up to the conclusion of the Games. Katniss narrates the events of, and her experience in, the Games in Chapters 11 through 25; and with the help of Peeta, the other tribute from District 12, Katniss wins the Games (Chapter 25).

A Note to Teachers

This chapter is appropriate for all levels of secondary mathematics as basic counting principles are necessary for most activities. The mathematics activities presented in this chapter follow the novel chronologically. As such, some of the latter activities are much less complex mathematically than those encountered earlier in the chapter.

BEFORE READING *THE HUNGER GAMES*

Time as a Resource

The availability of resources, both physical and social, is a recurrent theme in *The Hunger Games*. This activity, *Time as a Resource*, is intended to begin conversations around resources such as time and relationships and consider how these resources impact lives. Share the popular "motivational" saying, "You have the same 24 hours in a day as Beyoncé" (Brown, 2018). Ask students to write a short reflection about what they think the phrase means, how it makes them feel, and if they agree or disagree with the statement. After individually reflecting, encourage students to turn to a partner to discuss their thoughts. Conduct a short classroom discussion, highlighting varying interpretations of the phrase. One interpretation is that time can be considered an equalizer, that no matter the advantages or disadvantages of fame, we all have the same number of hours in the day. Other interpretations may include encouraging people to make the most of their time, to not put off tasks that will support your success, or that if Beyoncé can be a huge success, anyone else can too.

Students can then map out their "typical" daily life and imagine how that might compare to a celebrity. Have students write the hours from 6:00 a.m. to 10:00 p.m. down the left-hand side of the paper in 30-minute increments (see example in figure 11.1). Next, ask students to create three additional columns:

	EVENT	RELATIONSHIPS	TYPE
4 - 4:30	Soccer	Teammates	Wellness
4:30 - 5	Practice	Coaches	
5 - 5:30	Drive Home	Mom	Wellness
5:30 - 6	Homework	Alone	Work
6 - 6:30			
6:30 - 7	Help Cook	Dad & Sister	Work/Survival
7 - 7:30	Dinner	Family	Survival
7:30 - 8			
8 - 8:30	TV	Alone	Leisure
8:30 - 9			

Figure 11.1 Example Portion of a Student Daily Schedule. *Source:* Created by authors

Event, Relationships, and Type. Instruct students to first fill in their "typical" daily schedule under the Event column. This should include all activities that they do on most days such as cooking, eating, driving to and from different places, showering, school, chores, homework, sports, clubs, and any other frequent activity. Next, in the middle column, Relationships, ask students to identify the social interactions that are integral to that event. These can be family, friends, teachers, coaches, or no one. Finally, in the last column, Type, instruct students to create categories to group their Events into broad categories. Examples of types of Events can include survival (food), wellness (exercise, sports, religious practice, hygiene, etc.), work (school, after-school jobs, chores), or leisure (hobbies). It might help students to consider if each Event is optional for them.

Once students have completed their daily schedules, ask them what they notice about how they spend their days and their social interactions. Ask them to share their noticings with their peers, noting similarities and differences in their schedules. Lead a group discussion to identify themes in the classroom. Ask students to consider how this schedule may look for celebrities, by considering the following questions: *How might celebrities' schedules, social interactions, and categories look different from ours? What might they spend more or less time doing?* From this conversation, revisit the conversation about "having the same 24 hours in a day." *Do students think this is an accurate statement? How are their "24 hours" different from celebrities'?*

The people of Panem have vastly different "24 hours" depending on their district. Social status, wealth, community support, and resources vary for each district and determine how people spend their days. When she first arrives at the Capitol, Katniss considers,

> *What must it be like, I wonder, to live in a world where food appears at the press of a button? How would I spend the hours I now commit to combing the woods for sustenance if it were so easy to come by? (p. 65) (emphasis in original).*

Panem created a hierarchy of the "best" and "worst" districts, privileging some citizens to a life of luxury, while others had to scrounge for food or starve. However, despite their standing, the relationships and skills of each citizen are of great importance, and in Katniss's case, keep her alive. This activity prompts students to consider how they and the people in their community spend their time. While the lives of the elite may be idealized, the relationships central to our daily lives are deeply meaningful. This activity lays the groundwork for students to consider how their lives may be similar to the lives of the citizens of Panem and how the relationships prevalent throughout their days are a critical component of who they are and how they succeed.

Mathematical Probabilities

Probability, or the likelihood of an event (calculated as number of "desired" outcomes divided by total number of outcomes), will likely be a familiar topic for students, however, a review may be beneficial. Using the "Student Daily Schedule" previously created, students should have thirty-two 30-minute time slots on their schedules. Before reading the novel, ask them to calculate the probability of each "Type" of "Event" on their schedule. To do this, students should count up how many time slots are categorized as each "Type" and divide that number by 32 (the total number of time slots). If a time slot was selected at random, the probability is the likelihood that each "Event" would be selected. Have students consider which events have the greatest and smallest probability, and how that relates to their lives and how they spend their time. Students can also calculate the probability of being with people at any randomly chosen time by calculating their "Relationships" probabilities. This activity provides students with an introduction to calculating probabilities, which is necessary for the activities throughout the rest of the novel.

DURING READING *THE HUNGER GAMES*

The popular mantra from *The Hunger Games* (2008), "May the odds be ever in your favor," is tongue-in-cheek as the odds are stacked against not only the tributes, but all of the citizens of Panem that live outside of the Capitol (p. 19). To develop an understanding of the relationship between probability and odds, the class must develop a working definition of odds. The teacher can draw upon textual references like when Katniss describes how she and Gale have increased their odds by taking tessera to support their families (p. 13). The teacher can also draw upon this formal definition and examples to guide students' understanding:

> Definition of Odds: The probability that the event will occur divided by the probability that the event will not occur.
> Example 1: If a track athlete runs 100 races and wins 25 times and loses the other 75 times, the probability of winning is 25100 = 0.25 or 25%, but the odds of the athlete winning are 25750.333 or 1 win to 3 losses.
> Example 2: If the probability of the event occurring = 0.80, then the odds are 0.801 − 0.80= 0.800.20 = 4 (i.e., 4 to 1)

Once the class has developed working definitions of probability and odds, the students will be ready to apply these concepts in the following activities.

The Probabilities and Odds of Surviving the Hunger Games

The odds of survival in the Hunger Games depend on multiple factors. Given her expertise and sponsors' gifts, Katniss's odds for survival change based on what is valued by those in power. Students will use independent and dependent events to calculate the odds of survival within the Hunger Games while simultaneously considering how various traits and societal perceptions impact odds of survival.

Setting Up the Problem

As students read chapters leading up to the tribute training in Chapter 7, the class should discuss different survival skills afforded to different characters in the book. For example, Katniss has refined her tracking skills with Gale as they hunt outside of District 12 (Chapter 1). These discussions can be grounded in the experiences of specific characters, but also can be discussed around general skills afforded to tributes from different districts as a whole. For example, tributes that are from District 2 prepare their whole lives and often volunteer as tributes. These discussions should occur at the end of every chapter in preparation for this activity.

After reading Chapter 7, create a class list of the different survival skills that the tributes trained with during their initial training program. The list might include the following: spear, wrestling, ax, archery, knot tying, camouflage, making shelter, and identifying edible plants. Ask students to individually rank these skills based on their usefulness for survival before collectively agreeing on a classroom rank. During this process, ask students to justify their reasoning with references to the text. Students might draw upon Katniss's description of Career Tributes' thirst for the "deadliest-looking weapons" and how they "handle them with ease" (p. 95). Or, students might see a slingshot as having a lower probability due to Katniss questioning, "But what is a slingshot against a 220-pound male with a sword?" (p. 99). Once the class has ranked the list of survival skills, ask students to consider the variable of *friendship* within the Games. As the Hunger Games is essentially a type of reality television, friendship is integral to increasing attention and attracting potential sponsors. Students should take into account how the various tributes' characteristics from the different districts will be valued, or not, by sponsoring citizens. Based on their reading of Chapter 7, create a class ranked list of the 12 districts that reflects how valuable friendships would be for survival in the Hunger Games.

After creating the two lists, the teacher should assign probabilities of surviving the Games for each skill and friendship. Table 11.1 shows possible probabilities of 12 survival skills and 12 friendships. These probabilities are built on the assumption that if a tribute knew every survival skill they would have 100/100 of the skills category, and if they were friends with all districts they would have 100/100 of the friendship category. This table will be used to simulate probabilities, which will be used to calculate the odds of surviving the Hunger Games in the following two activities.

Odds Based on Survival Skills

Prior to exploring the combined odds of survival skills and friendship on surviving the Hunger Games, students must first calculate the odds of survival based on skills alone. For practicality, skills are discrete, meaning that they are not affected by one another. In other words, if a tribute knows how to use a knife, it has no impact on their ability to build a shelter. If a tribute knew every skill, theoretically their probability of winning the Hunger Games on survival skills alone would be 100,100. A tribute's total probability based on survival skills is calculated by adding up the probabilities of all skills. For example, if a tribute had the top and bottom ranked skill using the

Table 11.1 Potential Probabilities for Survival Skill and Friendship

	Highest --> Lowest										
Skill	30100	18100	10100	9100	8100	7100	6100	5100	4100	3100	2100
Friendship	40100	20100	10100	9100	6100	5100	4100	3100	1100	1100	1100

Source: Created by authors

probabilities in table 11.1, their probability of survival would be 30,100 + 2,100 = 32,100 = 0.32 on survival skills alone. Their odds of survival would be 321,001 − 32,100 = 3,210,068,100 = 3,268 = 817 or 8 to 17.

In partners, the students can participate in a simulation to determine their odds of surviving the Hunger Games based on randomly assigned skills. Each partnership will run one simulation together. This simulation uses cards numbered from 1 to 12. The first card a student draws is the number of survival skills they know. Then, with all cards face down, students select X random cards, where X is the number of skills determined by the first card flip. The newly turned over cards indicate which survival skills, from the class rank-ordered list, the "tribute" (student) knows. Since we assume skills are discrete, randomly turning over cards is justified. To calculate the odds of survival based on survival skill, first students must calculate the total probability of survival. To do this, they should add the probabilities from table 11.1 associated with the skills they turned over in the simulation. Once the students have calculated their probability of survival based on their drawn skills, they should also calculate their odds of survival. Teachers are encouraged to lead a class discussion on what it means to have odds less than one, equal to one, and greater than one and how this relates to probability and its boundaries of 0 and 1.

Impact of Friendship on Odds of Survival

As students continue to read, ask them to consider how some events are dependent on each other and how that can impact tributes' odds of survival. Present the following definitions:

> *Independent Event*: Two events are independent when the outcome of the first event does not influence the outcome of the second event.
> *Dependent Event*: Two events are dependent when the outcome of the first event influences the outcome of the second event.

After talking through these definitions, the teacher could lead a class discussion using the following question: *Are the probabilities of surviving the Hunger Games based on friendship independent of, or dependent on, the tributes' survival skills? Justify your answer with references to the book.* This discussion might surface connections to what society values within tributes, how districts might value different survival skills, and how sponsors might view friendships. The discussion should conclude with the class understanding that the probability of surviving the Games based on friendship is dependent on tributes' survival skills.

With their partner, students should have a discussion about which district's tributes would entertain a friendship with them based on the acquired survival skills and overall odds of survival calculated in the previous simulation. For example, if a partnership's simulated survival skills were wrestling, ax, and knot tying, which would lead to a high probability, thus high odds, this partnership will be more likely to be friends with the Career Tributes from Districts 1, 2, and 4. After deciding on which districts would be valuable friendships, students should calculate the probability of surviving the Games based on those friendships alone by adding up the probabilities of surviving based on friendship from the ranked list in table 11.1. Students should now have two different probabilities of surviving the Hunger Games, one based on survival skills and the second based on friendship—both of these should be out of 100.

Now it is time to calculate the odds of surviving the Hunger Games based on survival skills and friendship. To do so, students should calculate the final probability of surviving the Hunger Games based on survival skills and friendship by using the following formula:

$$P(Surviving\ the\ Hunger\ Games) = P(Survival\ Skills\ AND\ Friendship)$$
$$= P(Survival\ Skills)$$
$$\times P(Friendship\ Dependent\ on\ Survival\ Skills)$$

After calculating the final probability, the partners should now calculate their odds of surviving. Once the students have calculated their hypothetical final odds of survival, the teacher could facilitate a student-led discussion around their findings. Teachers should provide scaffolding during this conversation to help students make connections between how society's perceptions affect the odds of survival in the Hunger Games.

Exploring the Role of the Gamemaker, Theoretical and Experimental Probabilities, and Fair Events

In the previous activity, students explored the probability of tributes' survival during the Hunger Games based on internal and external factors, specifically their survival skills and friendships. An external factor not yet discussed is that of the Gamemakers. In this activity, students investigate the role of the Gamemaker, and more generally that of the "house" or "gamemaster." Students engage in a simulation to illustrate the impact a Gamemaker can have on the probability of survival in the Hunger Games. Begin this activity after reading Chapter 25 in the novel.

Simulation Activity 1: Is This Event Fair?

For this activity, each pair of students will need a physical or virtual die. Pose the question, "If the probability of an event is fair, is your chance of winning fair?," to the class for a brief orienting discussion. Students may think of games and events that are fair and compare those to unfair games. Games are unfair when every player does not have the same chance of winning. This is true for the second person to play in tic-tac-toe, or in games where specialized knowledge or skill increases the probability of winning, like a spelling bee or a science fair. A thorough discussion of fair and unfair will follow later in this activity and should be kept brief at this point. The intention of this discussion is to prompt students to think about what it means for a game to be fair.

Next, have students work in pairs to calculate theoretical and experimental probabilities.

Begin by telling students that "six" is the teacher's lucky number, and they need to pick their own lucky number between one and five for this activity. Students in pairs should not have the same lucky number. Students should first calculate the probability of getting their number in a single roll if the dice is fair, which is 1/6. Ask students to also calculate the probability of not getting their lucky number—that is, 5/6 or 83.3%—seemingly much more likely! Lead a discussion with students about how these calculations are theoretical probabilities, or what we would anticipate would happen in a "perfect" simulation. However, as students may anticipate, when they actually roll a die, the probabilities may differ from the theoretical probability. For example, theoretically if someone rolls a die 60 times, 10 rolls should be the teacher's lucky "6." Fifty of those 60 rolls should be divided equally between outcomes 1, 2, 3, 4, and 5. But, if we were to actually roll a die 60 times that might not be exactly what happens.

With a single die, allow the student pairs to "play" a few rounds, where they roll their die until they get their number, tracking how many rolls it takes to get their number. For instance if the student takes 10 rolls to get a 6, then their experimental probability of winning is 110. Their probability of losing is 910. For a challenge, have students track all of the outcomes of their rolls. Students can create a chart to keep track of rolls with tick marks to keep the experiment organized. Ask students to calculate the experimental probability of getting their lucky number with 1 roll, then 5 rolls, then 10, and so forth. Students may notice that with more rolls, their experimental probability moves closer to the theoretical probability they calculated earlier. This foreshadows the use of the Simulator 1.

Use a fair dice simulator (CODAP, 2021a). It is important that students are not aware of this. First, discuss the probability of rolling the teacher's lucky number (six) compared to anyone else's lucky number. Students should conclude that all outcomes have the same theoretical probability. To use the Simulator, it is necessary to get familiar with the controls. The simulation is set to roll once by clicking "Start." To clear data or restart the simulation, click the "CLEAR DATA" button. Play a few rounds, rolling the die once as a class. Ask the students their lucky number (other than six). Click "Start" to roll the die one time to see if the student, teacher, or no one wins. Then, change the number of rolls by adjusting the "samples" number, and simulate many rolls, to show that the experimental probabilities of rolling students' numbers and the teacher's number are about the same with many rolls. That is, the teacher and the student should win their games about the same number of times. For instance, in 60 games, the teacher's "6" should theoretically win about 10 games. The Simulator provides a graphical representation of the results, teachers are encouraged to discuss with the class what is collected within the representation by discussing the axes labels and what is represented by a data point.

After completing the first simulation, discuss the role of Gamemakers in *The Hunger Games*. The Gamemakers serve as antagonists in the novel—a group of residents of the Capitol who control all aspects of the game. Before the Games begin, the Gamemakers determine location, terrain, the Cornucopia, and more. During the game, the Gamemakers can change the course of play. For instance, to move players, they set a fire (Chapter 13). They can also change the rules, like from a single victor of the Games to shared or dual victors (Chapter 18). Additionally, the Gamemakers are responsible for the "production" of the Hunger Games; the Games are televised and some consider them entertainment, others a reporting of tragic events. The Gamemakers must consider their impact on the production value of the Games. That is, how entertaining and informative their production is for viewers.

Simulation Activity 2: How Could Probabilities of an Event Change from Fair to Unfair?

For the following activity, the second simulator (CODAP, 2021b), which is an unfair or loaded dice simulator (which should not be discussed with students at this time) will be used. Here, instead of each digit 1 to 6 having a ⅙ probability, 6 has very little chance (151) and all other digits, 1 through 5, each have a 1,051 chance of being rolled. As in Simulation 1, Simulation 2 is set to roll when "Start" is pushed; the number of rolls can change by adjusting the "samples" number, and "CLEAR DATA" restarts the simulation. Challenge students to play with the teacher, as they have in Simulation 1, with single rolls. Calculate and compare the experimental probabilities to the theoretical probabilities. The teacher will lose repeatedly. That is, in a single roll, six will rarely appear. In multiple rolls, the number of sixes will be significantly smaller than any other number the student may have selected to play and students will "win" frequently.

Discuss the fairness of Simulation 2. Draw connections to the fairness of the previous simulation. Obviously, the simulation is unfair since the theoretical probabilities are not equal for each outcome. For an inquiry extension, ask students how they might figure out how this die is unfair or loaded. To answer this, they may run multiple experiments and average the probabilities of each digit being rolled. Allow students to try varying ways of calculating the theoretical probabilities of your loaded dice before revealing the answer.

Follow Simulation 2 with a discussion of fairness and how the "unfair" simulation relates to the Gamemakers in the Hunger Games. That is, with the Gamemakers changing the game for the tributes as it is played, the probabilities of events happening change, but remain unfair between tributes. Students who play electronic games against a computer player may also draw from those experiences. Be cautious of the (sometimes nuanced) differences between topics that may have mathematically or statistically fair and unfair outcomes (e.g., games) and topics that are unfair due to unjust or unequal outcomes. Students should connect the role of the Gamemakers to the tributes' survival in the Hunger Games. For example, survival is impacted by survival skills and friendship. However, regardless of these factors, the Gamemakers can change those probabilities in many ways. Discuss why Gamemakers might make these choices and how this might be connected to the perceived entertainment and production value of the Hunger Games. For example, the Gamemakers may not provide an ax for a tribute who has that particular skill. This tribute may talk about their disappointment in not receiving an ax, or their lack of training or expertise with the tools they do have—which may make viewing more interesting if it's unclear whether the tribute will be successful.

It is in the Capitol's interest to have the "odds" (or probabilities!) forever in their favor, so they mostly determine the outcome of the Games. Katniss is keenly aware of this. Many of her decisions are impacted by what she believes the Capitol, viewers, and Gamemakers want to see. While the Games are unfair, ask students to consider how Katniss uses this knowledge as a form of power. For instance, in Chapter 19, when starving and hoping Haymitch would send food, Katniss decided to kiss the injured Peeta. She reflected on how Haymitch sending food showed her that the viewers would want to see a romance between her and Peeta. This scenario is repeated in Chapter 22.

By Chapter 26, Katniss and Peeta have forced the hand of the Head Gamemaster to declare them co-victors. Have students discuss how this rule change (Chapter 18), which was at one time reversed (Chapter 25), impacted their (theoretical) probabilities for survival. Upon the initial rule change, Katniss rejoined efforts with Peeta. Their time together allowed them to discuss game strategy, sharing what each knew and the other did not about the Games to combine their efforts, like when each is injured or tired at different times and the other takes care of them both (Chapters 19 and 22) and sleeping in shifts (Chapter 23). Teachers can conclude this discussion by asking students to consider what events they think should be fair, but are not. Potential topics of conversation include school funding, graduation rates, elections, income, health care, housing, and more. For many of these topics, there are "gamemakers" and the rules keep changing. Some prompts that a teacher can use to guide this discussion and encourage students to make predictions can be the following: *What must be considered or included in any calculation of a tribute's probability of winning? Tributes do not have the same probability of winning. Why? Is this "fair"?* and *"Would the world be more "fair" if all people had the same probability for success (or desired outcomes)? Explain why or why not.*

AFTER READING *THE HUNGER GAMES*

Choices versus Chances

This final activity acts as an assessment of students' conceptual understanding of conditional probability as applied to Katniss's experiences in *The Hunger Games* novel. A goal of this activity is for students to notice how one single event or choice may not control all outcomes, but these events or choices in combination with other events may increase the probabilities of good outcomes.

Introduce the activity by reminding students that they observed in the while-reading activity that the probability of many events are conditional upon others since survival based on friendship was dependent on survival skills. Additionally, in "Simulation Activity 2: How Could Probabilities of an event Change from Fair to Unfair?," students observed that not all outcomes have equal probabilities of occurring. Because of the interdependence of events and the reality that not all occurrences are mathematically fair, it can be difficult to discern the numerical probabilities of events occurring. This leads to the question: If it is difficult to know the probability of any one event happening, how can we use principles of probability to make choices that will end in outcomes that are favorable to us?

Ask students to reflect back on the events of the book, the personality traits of the characters, and choices of the characters that they believed played a major role in the fates of Katniss and Peeta in the Games. Ask students to choose one of these factors as having the biggest impact on the outcome of the Games. We suggest putting some limitations on choices to ensure some expansion in the discussion; for example, it is probably best to exclude the moment when Katniss and Peeta threaten to eat the poisonous berries as that moment is so near the end of the Games. Students can choose a single event, a character decision, or a characteristic of the characters. Some examples may be Peeta's love of Katniss, Katniss shooting the pig, or Katniss choosing to ally with Rue. Once students choose their *significant event*, they will develop a complete argument for why their event was integral to the outcome of the Games. This argument will also address the following questions: How did the significant event increase the likelihood of other events occurring or not occurring? If that event did not occur, what other events would be more or less likely to happen? What prior events or situations led to the occurrence of the significant event?

Students will compose their positions into a short, persuasive commentary similar to those on popular websites like an SR Originals (Screen Rant, n.d.). Students can choose to create their commentaries as a blog post or a FlipGrid video. Students will read or view each other's commentaries. To facilitate debate or discussion of the posts, ask students to comment on the persuasiveness of their peers' arguments with respect to their attention to how other events were conditional upon the factor argued to be significant.

In a whole-class discussion, ask students what they noticed about the events chosen by their classmates. You may wish to categorize the events discussed. Were the events-conscious choices made by the characters, were they aspects of the characters' personalities, or were they events outside the characters' control? In keeping with the themes of the novel, it is probable the events students identify are ways the characters acted to either be true to themselves, to show compassion or loyalty, or to work creatively within the existing oppressive structures to better their chances of survival. Some significant events may be the result of the creation of friendships and allies inside and outside the Games. Revisit the guiding question of the activity: If it is difficult to know the probability of any one event happening, how can we use principles of probability to make choices that will end in outcomes that are favorable to us? Engage in discussion of how the commentary activities provide answers to the question.

Conclude the discussion by having students reflect on their own lives. Ask students to list on their papers some aspects about their lives or opportunities outside of their control (e.g., where they live, their responsibilities, their age). Ask students to write down a goal they have for themselves. Do they see these aspects outside their control inhibiting their chances of attaining a goal? Remind students of the choices Katniss or other characters made to better their chances of survival despite the events that were outside their control. Give students individual time to brainstorm and journal about choices, connections to others (community, family, friends), and personal development that will help increase their chances of reaching their goals despite the factors they acknowledge are outside their control. Optionally, have students create a display (e.g., graphic, picture, concept map) of the ways that these connections, choices, and plans will help them reach their goals within the constraints of their existing situations.

EXTENSION ACTIVITIES

In the culture of Panem, the "odds" favored certain people or communities over others solely based on geographic and societal structures. Katniss demonstrated that sometimes individuals and groups of individuals can act to better their own odds or even dismantle the systems which perpetuate these inequities. In the United States, people can bring such change through voting for leaders and legislation. The following activities provide opportunities for students to ponder and investigate how the fictional inequalities in Panem relate to their own communities and how different communities afford varying conditions for voting.

Introduction to the Simulation

Explain to students that while voting is a constitutional right granted to U.S. citizens, there are rules and systemic protocols states and individuals must follow to have their votes counted. These rules vary by state and region and may affect certain individuals more than others. Explain some of the processes and considerations for voting eligibility including needing a government-issued photo ID, being able to register on Election Day, having time to travel to a polling place and wait in line to vote, and the possibility that one's name is removed from the list of registered voters by mistake. Before launching into the simulation, facilitate a class discussion regarding what they believe the probability that an eligible voter who wants to vote is able to vote.

Will You Get to Vote?

Students will participate in a simulation where the rolling of dice will determine their likelihood of being able to vote. Steps for the simulation are provided at the end of this section. The teacher may choose to have students complete the simulation in groups or lead the whole class to complete the steps together. Prior to the simulation, review the probabilities of outcomes from rolling one or two six-sided dice utilizing the discussion from Simulation Activity 1 as well as the sample space for rolling two six-sided dice provided in figure 11.2. The discussion of the probabilities of the six-sided dice may include determining the total number of outcomes (36) and then the probability of rolling a certain outcome like 7 (6/36 since 7 can occur in 6 ways).

This simulation takes place the night before Election Day and every student is a U.S. citizen legally eligible and motivated to vote. For the purposes of the

		Die 1 roll					
		1	2	3	4	5	6
Die 2 roll	1	2	3	4	5	6	7
	2	3	4	5	6	7	8
	3	4	5	6	7	8	9
	4	5	6	7	8	9	10
	5	6	7	8	9	10	11
	6	7	8	9	10	11	12

Figure 11.2 Sample Space for Rolling Two Six-Sided Dice. *Source*: Created by authors

simulation, we assume that if the voter has the government identification, they have a preexisting voter registration. Each students' ability to vote will be determined by the following factors: state of residency, possession of government-issued photo ID, nearness to polling location, and work and home responsibility schedule. Students will roll one die to simulate which of six states they live in. Data in the State Table are taken from published regulations (National Conference of State Legislatures, 2020a; 2020b). The teacher may choose to substitute or add states to modify the table. For example, if 2 dice are used, the class could select from 11 states. What follows is the Voting Simulation instructions.

1. **What state do you live in?** Roll <u>one</u> six-sided die. Your outcome corresponds to the state you are voting in according to the State Table below.

State Table

Dice Roll	A: State	B: Photo ID Required?	C: Same Day Registration?	D: Voting Format	E: Time off to vote?	F: Polling Hours
1	Colorado	No	Yes	Mail in	Depends	7am - 7pm
2	Missouri	No	No	In person	Depends	6am - 7pm
3	Georgia	Yes	No	In person	Depends	7am - 7pm
4	Florida	No	No	In person	NO	7am - 7pm
5	California	No	Yes	In person	YES	7am - 8pm
6	Wisconsin	Yes	Yes	In person	YES	7am - 8pm

2. **Do you have sufficient identification for your vote to be validated?** Roll <u>two</u> dice. If you did not roll a 2, then you have an official photo ID. Move on to Step 3. If you rolled a 2, then you do NOT have an official photo ID. Check column B in the State Table above to determine if your state requires official ID.
    o If NO, move on to Step 3.
    o If YES, your vote will not be able to be validated. **Your vote won't count so you won't vote. Go to Step 7.**
3. **Is your voter's registration still in the system?** Roll <u>two</u> dice. If your outcome was not 12, then your registration is still in the system. Move on to Step 4. If your outcome was 12, then your name has been purged from the registration database due to system maintenance. Check the State Table column C to see if your state will allow you to register on election day.
    o If YES, move on to Step 5 since you will need to register and vote in person.
    o If NO, you are not able to register to vote. **You cannot vote. Go to Step 7.**
4. **Will you need to vote in person on election day?** Check the State Table column D. Does your state do MAIL IN voting or IN PERSON?
    o If MAIL IN, you received your ballot two weeks ago and have already returned it. **You get to vote! Go to Step 7.**
    o IF IN PERSON, move on to Step 5.
5. **Can you go vote during work hours?** Roll one die to determine your work and home responsibility hours using the table below.

Work & Home Responsibility Hours

Die roll	1	2	3	4	5	6
Work Hours	7:30a - 4p	9a - 5p	7a - 7p	6a - 2p 3:30p - 8p	8a - 3p	8a - 5p

| Home Responsibility Hours | 5p - 8p | -- | -- | -- | 3:45p - 7p | 6a - 7a
6p - p |

- Compare your WORK hours to your State's polling hours (column F of the State Table) and time off (column E) to determine if your employer must give you time off to go vote.
  - If YES, you can leave work to go vote. **You get to vote! Go to Step 7.**
  - If NO, proceed to Step 6.
  - If DEPENDS, compare your WORK hours to your State polling hours. Is there a three-hour chunk of time that the polls are open while you are NOT at work?
    - If YES, then your employer will not give you time off to vote. Proceed to Step 6.
    - If NO, then you can leave work to go vote. **You get to vote! Go to Step 7.**
6. You will need to vote before or after work. Do you have the time?
   - Roll <u>one</u> die to determine the time to travel to your polling location.

Travel time to polling place			
Die Roll	1 or 2	3 or 4	5 or 6
Travel time from your work or home	10 min	15 min	20 min

   - Roll <u>one</u> die to determine the amount of time you will need to wait in line to vote.

Wait time at the polls						
Die roll	1	2	3	4	5	6
Wait times at the polls	5 min	8 min	10 min	15 min	25 min	45 min

   - Analyze if you have time to vote. Things to consider:
     - ☐ You need to arrive at the polling place before it closes.
     - ☐ Assume the polling place is the same distance to your home as your to your work.
     - ☐ You need to get to your work or home responsibilities on time. You might need to consider travel time to and from the polling place.
7. Did you get to vote? _____

Calculation of Experimental and Theoretical Probability

Lead a discussion on the experimental and theoretical probability of any one person in the class being able to vote in this simulation by first asking what probability of being able to vote they believe to be fair. We expect that students will mention an eligible voter who wants to vote should have close to 100% probability of being able to vote. The teacher might ask how much less than 100% would be acceptable to the students. Next, compile the class data by tallying how many students were able to vote and how many were not. Use these numbers to determine the experimental probability of being able to vote in this simulation. The theoretical probability of being able to vote in each state is given by the compound probability:

$$P(\text{getting a vote}) = P(\text{Having sufficient ID})$$

$$\times P(\text{Having registration given sufficient ID})$$

$$\times P(\text{Having time to vote given having registration})$$

Having the time to vote is conditional on multiple factors in the simulation including having sufficient ID, having a registration, travel time to polling locations, waiting times, and personal time commitments. We suggest listing the possible outcomes table for one state and demonstrating the probabilities there (see figure 11.3).

Discussion of Factors that Affect Ability to Vote

The goals of this discussion are for students to (1) identify policies that make voting more or less accessible, (2) identify populations of people that may be disproportionately affected by these policies, and (3) connect their observations about (1) and (2) to take a stance on the policies they would like to exist in their communities' voting legislation.

Lead students to discuss what they noticed helped or hindered their ability to vote. The following questions may generate a lively discussion: *What factors made voting easier? What made voting harder? Why might these "harder" factors exist? Why might some states not choose to implement the factors that make it easier? How do these factors relate to the differing conditions in Panem?* Record some of the students' noticings so the class can refer back to these ideas later in the activity.

The citizens of the Districts of Panem outside the Capitol were not able to vote for the laws and leaders that would govern Panem society. In the United States, citizens in all states are eligible to vote. However, as seen in the simulation, not all eligible voters indeed can vote. The possibilities in the Voting Simulation are randomly distributed possibilities. While all these possibilities exist in the real world, they are not equally or randomly distributed. For example, it has been found that wait times in predominantly non-white polling districts are longer than those in predominantly white neighborhoods (Fowler, 2020). Ask students to envision or brainstorm who may be more impacted by these circumstances by posing some of these questions: *Describe an instance where someone would have to travel longer to get to a polling location. List a person or profession that has each of the six job and home responsibilities. How long are polling wait times in our town? Where have*

In Georgia, if you do not have a photo ID, you are not able to vote. In addition, if your name is purged from the voting lists, you will not be able to re-register in person on election day. Therefore, in this simulation the probability of having sufficient ID is 3536, and the probability of having a voter's registration is 3536.

In the table below, we see there are eighteen possibilities for the lengths of voting and traveling. We split the table to indicate for some jobs, individuals need to consider two-way travel.

		Wait times at the polls					
		5 min	8 min	10 min	15 min	25 min	45 min
One-way travel times to the polls	10 min	15	18	20	25	35	55
	15 min	20	23	25	30	40	60
	20 min	25	28	30	35	45	65
Two-way travel times to and from the polls	20 min	25	28	30	35	45	65
	30 min	35	38	40	45	55	75
	40 min	45	48	50	55	65	85

Those with Jobs 2, 3, and 6 can vote during work. A person with Job 1 has between 4pm and 5pm to travel to the polls, wait in line, and travel home; 14 of the 18 time possibilities meet this. Someone with Job 4 has 90 minutes to travel to the polls, wait in line, and travel to their second job; all 18 possibilities are less than 90 minutes. A person with Job 5 can vote at 7am if they can vote and travel one way to work by 8am, 17 of the 18 times will meet this.

$$P(\text{Having time to vote}) = 36 + (161418) + (161818) + (16 \times 1718) = 103108$$

$$P(\text{able to vote in GA}) = P(\text{having ID}) \times P(\text{not purged}) \times P(\text{having time to vote})$$

$$P(\text{Being able to vote in GA}) = 35363536 \times 103108 = 126{,}175139{,}968 = 90.1\%$$

Figure 11.3 Example of Probability of being able to Vote in Georgia. *Source*: Created by authors

you heard of the lines being long? What are some reasons someone might not have a government-issued ID? In which kinds of jobs do you think it is more likely for people to get time off to go vote? What kinds of jobs are less likely people would get time to go vote? To culminate this activity, students will investigate one factor that they believe has an influence on an individual's likelihood of being able to vote or a factor that they believe disproportionately affects certain populations of people. Students will research the laws and conditions of that factor in their local community and create a product (presentation, letter, essay, etc.) to persuade community members or state legislators to either maintain the current conditions and laws or to change them.

CONCLUSION

This chapter proposed a probabilistic investigation of the odds of success in life based on personal choices and societal expectations. Through the reading of *The Hunger Games*, students interrogated what is meant by success, how certain personal characteristics help someone achieve success, and in what ways the definition of success is manipulated by societal forces. Activities guided students in the development of their mathematical literacy by interrogating the ways in which society manipulates its citizens through expected norms and values. As such, enactment of these tasks can draw on a critical stance as students discuss what the society of Panem finds valuable and not valuable within *The Hunger Games*. When engaging with these mathematical activities in conjunction with reading the novel, students have the opportunity to explore ways that their own lives are grounded in relationships that hold a greater importance than might be valued by society.

REFERENCES

Brown, D. (2018, August 24). *Stop believing you have the same 24 hours as Beyonce.* https://www.inc.com/damon-brown/stop-believing-you-have-same-24-hours-as-beyonce.html.

CODAP. (2021a). *Dice simulator #1.* https://bit.ly/3t5Q2j5.

CODAP. (2021b). *Dice simulator #2.* https://bit.ly/3uEjBsu.

Collins, S. (2010). *The hunger games.* Scholastic Press.

Fowler, S. (2020, October 17). *Why do nonwhite Georgia voters have to wait in line for hours? Too few polling places.* Georgia Public Radio. https://www.npr.org/2020/10/17/924527679/why-do-nonwhite-georgia-voters-have-to-wait-in-line-for-hours-too-few-polling-pl.

Hajnal, Z., Kuk, J., & Lajevardi, N. (2018, July 26–27). *A disproportionate burden: Strict voter identification laws and minority turnout* [Paper presentation]. The 2018 Election Sciences, Reform, and Administration conference, University of Wisconsin-Madison. https://esra.wisc.edu/wp-content/uploads/sites/1556/2020/11/hajnal.pdf.

National Conference of State Legislatures. (2020, August 25). *Voter identification requirements | Voter ID laws.* NCSL. https://www.ncsl.org/research/elections-and-campaigns/voter-id.aspx.

National Conference of State Legislatures. (2020, October 6). *Same day voter registration.* NCSL. https://www.ncsl.org/research/elections-and-campaigns/same-day-registration.aspx.

Oxford Dictionary. (n.d.). Citation. In *Oxford Dictionary* on lexico.com. https://www.lexico.com/en/definition/social_capital.

Screen Rant. (n.d.). *SR originals: The best movie and TV editorials, features and lists from the entertainment experts at Screen Rant.* https://screenrant.com/sr-originals/.

Subject Index

agency, xi, 43, 50, 130
algebra, xi, 37, 87, 99, 101, 104, 106, 127, 129, 132, 150, 157
algorithms, xii, 1, 26, 175, 185–88
app design, 186

binomial(s), 132, 143
book walk, 64, 65, 121
budget, 25, 28, 36, 41, 48

character(s), ix, xii, 7, 8, 18–20, 25, 26, 28, 30, 33, 35, 37, 41, 43, 50, 54, 55, 63, 64, 79, 81, 84–87, 90, 92, 93, 95, 97, 103, 104, 114, 117–23, 126, 127, 129, 131–36, 138, 140, 144–52, 173, 175, 180, 184, 185, 195, 202, 203; character arcs, 131, 138, 145, 148; characterization, xi, 79, 86, 117, 122, 124, 125, 127, 129, 145, 148; character tracking, 138, 145, 147
conflict(s), xi, 25, 28–30, 33, 34, 36, 38, 39, 41, 85, 93, 102
coordinate plane(s), 117, 120, 122, 124–27, 130
critical literacy lens, 43, 44, 60
culture(s), xi, 6–8, 99, 131–33, 136, 152, 176, 203; Japanese, 131–34, 136, 137; Vietnamese, 2, 6–8, 19

debate(s), 74, 83, 103, 104, 203
distance(s), 16, 31, 32, 54, 55, 63, 67–70, 124, 143
dystopia(s), xii, 99, 101, 102, 113, 114

events, xii, 63–64, 75, 85, 93, 102, 106, 113, 114, 126, 146, 147, 179, 181, 192, 194, 195, 197–203; dependent, xii, 195, 197; independent, xii, 195, 197

fairy tales, 2, 7, 8, 18
financial literacy, xi, 39, 40
fractions, xi, 2, 5, 8, 11–14, 21, 27, 55
functions, xi, xii, 99, 110, 111, 131, 132, 139, 141, 143, 144, 181; concepts, 109; interpreting, xi, 99; notation, 109
funds of knowledge, 1, 2, 15, 16, 21

graph, 105–8, 111, 113, 181, 182; graphing, 111, 127, 181; graphs, 52, 105, 108, 109, 111, 113, 181
graphic novel, xi, 1–2, 5, 9, 17–21

inequalities, xi, 99, 104, 107, 108, 113, 204

Subject Index

map(s), 63, 65–68, 70, 76, 81, 110, 118, 119, 121, 150, 168, 192, 203; concept map(s), 81, 168, 203; mapping, 66, 70, 75, 124; story map, 150
matchmaking, 175, 176, 180, 184–88
mathematics, ix–xii, 1–3, 6, 8, 9, 11, 14–22, 25, 26, 31, 37, 41, 43, 44, 48, 50, 54, 55, 60, 63, 68, 74, 79–81, 83–85, 90, 97, 99–104, 110, 113, 114, 117, 122, 128, 129, 131–33, 138, 140, 143, 144, 146–48, 150, 152, 155–58, 161, 164, 171–73, 175, 181, 187, 188, 191, 192; attitudes towards, 131, 138, 144, 146; identities, 50
mean, 143
measurement, 18, 43, 51–53, 93
mental mathematics, xi, 25, 26, 41

number sense, 156, 157

odds, xii, 194–98, 201, 204, 209

paratext, 45
partition numbers, 150
power dynamics, 44, 46, 48, 50, 56
precalculus, xii, 131, 132, 173
prediction(s), 67, 68, 81, 85, 124, 134, 139, 161, 184, 185, 202

probability, xi, xii, 46–48, 50, 60, 117, 121, 122, 125–27, 130, 175, 177, 179–88, 194–200, 202–4, 206, 207; experimental, xii, 198–200, 206
proportionality, 53
Pythagorean Theorem, xii, 117, 118, 120, 123, 125, 127, 128, 130, 158, 160

rational number(s), xi, 2, 3, 9, 11, 14, 15
ratios, 50, 51, 55, 103, 107
refugees, 63, 71–73

safe water, xi, 64, 75–77
scale, xi, 3, 21, 29, 30, 63–76, 79, 81, 183
sequences, 25, 28, 112, 131, 133, 138, 139, 141–44; arithmetic, 112, 133; geometric, 112, 133
series, xii, 32, 33, 46, 59, 64, 85, 114, 131, 132, 135, 146, 167
speed, 52, 68, 69, 143
storytelling, 7, 134, 173

theoretical probability, 198–200, 206, 207
translated text(s), 131, 152

unknown(s), 113, 117, 118, 124

variable(s), xii, 47, 48, 68, 105, 107, 113, 175–77, 180, 184–86, 188
visual literacy, xi, 5, 19

About the Editors

Paula Greathouse is an associate professor of Secondary English Education at Tennessee Tech University. She has coedited several books, including *Adolescent Literature as a Complement to the Content Areas* (2017) book series, and coauthored *Developing Adolescent Literacy in the Online Classroom: Strategies for All Content Areas*, which won the 2021 Divergent Book Award for Excellence in 21st Century Literacies Research. Her research on adolescent literacy and young adult literature has been published in books and top-tier journals such as *Educational Action Research, Study and Scrutiny: Research on Young Adult Literature, The Clearing House,* and *English Journal*. She was a secondary English Language Arts and Reading educator for 16 years. She has received several teaching awards including the National Council of Teachers of English (NCTE) Teacher of Excellence. She is an active member of the NCTE, International Literacy Association (ILA), Association of Middle Level Education (AMLE), and the Assembly of Literature for Adolescents of NCTE (ALAN).

Holly Garrett Anthony is professor of mathematics education at Tennessee Tech University. She is responsible for teaching mathematics content and methods courses for preservice teachers, as well as doctoral courses in Science, Technology, Engineering, and Mathematics (STEM) education, qualitative research, and program planning. Her research focuses on the professional development of teachers and STEM education. Her work on communication in the sciences, improving clinical experiences, and responding to social injustice has been published in 11 book chapters and several top-tier journals including *Mathematics Teacher* and the *Journal of Multicultural Affairs*. She has previously taught mathematics in public and private middle and high schools in two states and three countries. She is an active member of the National Council of Teachers of Mathematics (NCTM), the Association for Mathematics Teacher Educators (AMTE), and the International Congress of Qualitative Inquiry (ICQI). She also serves in leadership roles in three state mathematics education organizations.

About the Contributors

Robin Keturah Anderson is a former middle and high school mathematics and physics teacher from Southern California. She is currently an assistant professor of mathematics education at North Carolina State University where she teaches courses to preservice and in-service mathematics teachers with the goal of promoting anti-racist pedagogies. Her research interest focuses on nontraditional forms of professional development through teacher-initiated learning in social media.

Carlos Castillo-Garsow is an associate professor of mathematics education in the Mathematics Department at Eastern Washington University. Dr. Castillo-Garsow comes from a background in applied mathematics and linguistics, and currently studies how students and mathematicians use mathematics as a language to express their ideas. He primarily teaches mathematics content and methods courses for preservice mathematics teachers, with a focus on middle and secondary mathematics.

Katherine Baker is an assistant professor at Elon University and supports prospective teachers in their mathematics education and instruction. She collaborates with prospective and practicing teachers and teacher educators to explore students' mathematical thinking and how to use that thinking in instructional decision-making.

Rachel Colby is a high school mathematics teacher who has taught Algebra 1, Algebra 2, and Geometry for four years. She also taught middle school math for four years before transitioning to high school. Rachel holds a Bachelor of Science in Middle Secondary Education and Mathematics from Butler University and is completing a Master of Arts in Mathematics with a Teaching Emphasis from the University of Northern Colorado. As an undergraduate, she researched incorporating literature in a mathematics classroom and has continued to incorporate literature in her classroom as a practicing teacher. Rachel also spent a year teaching English in France and was even able to incorporate literature in her math classes there.

Mike P. Cook is an associate professor of English Education at Auburn University, where he teaches undergraduate and graduate courses within the English Education program. His research interests include multimodal literacy and teacher identity development, including the ways teachers develop as critical practitioners and activists. His scholarship has appeared in *Journal of Teacher Education*, *English Teaching: Practice & Critique*, *Journal of Language and Literacy Education*, *ALAN Review*, and the *Journal of College Literacy and Learning*, among others.

Jennifer Edelman is an associate professor of mathematics education in the Early Childhood through Secondary Education department at the University of West Georgia. Building on her successful career as a public school teacher in K-8 education, she earned a PhD in Curriculum and Instruction: mathematics education from the University of Wyoming in 2014. In her work at UWG, she teaches classes on mathematics pedagogy and assessment, teaching methods and materials in Language Arts, and works with teacher leaders in courses on using data to meet the needs of diverse learners. Dr. Edelman's research examines how mathematics teachers learn to use curricular materials such as manipulative, lesson plans and children's literature in teaching mathematics.

Bryan Fede is a former high school mathematics teacher of eight years who has widened his interest in studying teachers' pedagogical practice from traditional face-to-face environments to classrooms on the digital landscape. He received his PhD from The University of North Carolina at Chapel Hill and currently serves as an Instructional Designer in the Division of Digital Learning at Marquette University.

Shelly Furuness is an associate professor of Education in Butler University's College of Education in Indianapolis, Indiana. At the undergraduate level, Dr. Furuness teaches face-to-face, hybrid, and online courses related to foundational learning theory, adolescent literature, content-specific literacy skills for middle secondary classrooms, methods, and curriculum design. At the graduate level, she teaches introductory courses on research particularly geared toward qualitative designs to investigate personally relevant inquiries, and she conducts lots of independent studies with graduate students interested in solving curricular dilemmas and problems of practice. Dr. Furuness is very interested in systems thinking, developmentally appropriate practices for adolescent and adult learners, teacher socialization, identity development, critical theory, and dialogic and relationship-oriented pedagogy.

About the Contributors

Rebecca Grice Gault is an assistant professor of mathematics education in the Early Childhood through Secondary Education department at the University of West Georgia. After teaching at the middle school and high school level, she returned to the University of Central Florida to obtain her PhD in mathematics education. Dr Gault's interests include implementation of cognitively challenging tasks in K-12 classrooms, integrating content areas including mathematics and literacy, and students' ways of reasoning about mathematics.

Kaylee Gentry is a graduate student in the Library Science Master's program at Tennessee Tech University. She has served as a teaching assistant in both young adult literature (YAL) and adolescent literacy courses in the College of Education's Department of Curriculum and Instruction. She is an avid reader of YAL and is working toward the goal of becoming a school librarian where she can help develop the love of reading in as many students as possible.

Allen Harbaugh is a learning scientist and statistician. He is an international scholar, and his work focuses on epistemic beliefs and how to impact student motivation and performance in statistics and mathematics classrooms. Currently, Dr. Harbaugh is an assistant professor of Statistics at Longwood University.

Amanda Huffman is a secondary mathematics teacher who has taught Algebra II, PreCalculus, Calculus, and International Baccalaureate mathematics for nine years. In addition to teaching, she has worked with numerous preservice mathematics teachers through providing mathematics methods instruction both in person and virtually, as well as hosting practicum students and student teachers. Amanda holds a Bachelor of Science in Middle Secondary Education and Mathematics and a Master of Science in Effective Teaching and Leadership both of Butler University. As an undergraduate, she researched integrating literature and writing in a mathematics classroom. She is also working on a Graduate Mathematics Certificate from Indiana University and has begun taking classes in pursuit of a PhD in Curriculum and Instruction with a focus in mathematics education from Purdue University.

Candace Joswick is a former middle and high school mathematics teacher in Florida and Ohio. She is currently an assistant professor of mathematics education at the University of Texas at Arlington where she is the program coordinator for Science, Technology, Engineering, and Mathematics (STEM) Education and the Master of Education programs for practicing PreK-12 mathematics and science teachers. Her research focus is on learning progressions and trajectories, technology, and classroom interactions.

Jen McConnel is a teacher educator and writer. She is a former English Language Arts teacher and her work focuses on supporting teachers and students across contexts. Currently, Dr. McConnel is an assistant professor of English Education at Longwood University.

Jennifer R. Meadows is an assistant professor at Tennessee Tech University in the College of Education where she teaches elementary math methods and graduate courses in STEM education. Prior to attaining a PhD in Exceptional Learning with a STEM education concentration and transitioning to teaching in higher education, she served as an elementary teacher for 14 years in Putnam County, where she was selected as the teacher of the year in 2012. Her research focuses on mathematics teaching and learning, preservice teacher education, and informal STEM learning.

Suki Jones Mozenter learned to be a teacher from the diverse and multilingual communities in Oakland, California. Now, she learns from the teacher candidates in the Integrated Elementary and Special Education program at University of Minnesota Duluth where she is an assistant professor of Reading/Literacies. Her research focuses on how students make sense of themselves as readers and writers, anti-racist pedagogy, and transformative teacher education.

Summer Melody Pennell is a Lecturer in education at the University of Vermont. Her research interests include YAL, social justice pedagogy, queer theory and pedagogy, and teacher education. She sometimes writes about interdisciplinary math education with Katherine and Bryan.

Brea C. Ratliff has taught mathematics at the elementary, middle, high school, and collegiate levels; been an academic coach for mathematics and science; been Master Math Teacher; and been a district K-12 mathematics supervisor. Brea is currently pursuing a PhD in secondary mathematics education at Auburn University, where she has been recognized as a recipient of the Presidential Graduate Opportunity Program fellowship. Her research interests include equity in mathematics education, teacher education and leadership, and community/youth development. She is the immediate past president of the Benjamin Banneker Association, Inc (BBA, a national organization established to advocate for high quality mathematics instruction for all students, with an emphasis on students of African ancestry. She is also the founder and educational strategist of Me to the Power of Three, a consulting company which specializes in developing inquiry-based programs and resources to support the teaching and learning of mathematics at all levels.

Jenna Repkin is a senior at Butler University studying Mathematics and Secondary Education. Jenna has experience in classes and conducting research in areas including General Literacy, Mathematical Literacy, and Methods for Teaching Multilingual Learners. Jenna has put together text sets and lesson plans in order to cater to student's developmentally appropriate literacy needs, thus engaging them further in learning. As a preservice teacher, Jenna has experience teaching and tutoring middle school math, Algebra 1, PreCalculus, and Calculus 2. She is getting ready to teach students in a High School PreCalculus class, and plans to continue teaching High School math after she graduates.

Brian Rothbaum spent three years teaching middle school in Florida, and has since been teaching high school in the Boston Public Schools for more than 20 years. He has taught everything from Pre-Algebra to Calculus to freshman through seniors. He received a Bachelor of Arts in Psychology with a minor in Math from Quinnipiac University, a Master's in Special Education Framingham State University, and a Master's in Counseling with a focus on Guidance from Cambridge College. Originally from New Jersey, he now resides in a suburb of Boston with his wife and three kids.

Shelly Shaffer is an assistant professor of Literacy in the Department of Education at Eastern Washington University, Cheney, Washington State, where she teaches preservice elementary and secondary teachers. She has taught content area literacy and writing, secondary methods, several Young Adult and Children's Literature courses, and graduate research courses. Her current research interests are YAL, reading motivation, flipped classrooms, mentoring, and multimedia integration in teacher education programs.

Lisa Skultety is a former middle and high school mathematics teacher in Houston, Texas, and is currently an assistant professor of Mathematics at the University of Central Arkansas. She teaches elementary and secondary mathematics education and researches ways to support teachers' noticing of students' mathematical thinking in the classroom.

Amber Spears is an assistant professor at Tennessee Tech University in the College of Education where she teaches elementary methods and graduate courses in literacy. She is a former elementary school teacher and is licensed in Tennessee in K-8 elementary education, PK-3 Early Childhood education, and PK-12 reading specialist. She is the chair of the Upper Cumberland Literacy Association and spends her summers facilitating literacy programming for young children in her community.

Marilyn E. Strutchens is an Emily R. and Gerald S. Leischuck Endowed professor and a Mildred Cheshire Fraley Distinguished professor of mathematics education in the Department of Curriculum and Teaching at Auburn University, where she is currently serving as acting department head. She teaches undergraduate and graduate mathematics education courses. She is the leader of the Clinical Experiences Research Action Cluster for the Mathematics Teacher Education Partnership. Her other research interests include equity in mathematics education, teacher leadership, and reform mathematics professional development for grades K-12 teachers. She serves as the chair of the Advisory Committee for the Directorate of Education and Human Resources for the National Science Foundation. She is a past member of the Board of Directors for the National Council of Teachers of Mathematics and a past president of the Association of Mathematics Teacher Educators, and a member of several national organizations' committees and writing groups.

Melissa Troudt is a former high school mathematics teacher in Colorado. She is currently an assistant professor of mathematics education at the University of Wisconsin—Eau Claire—where she teaches mathematics courses for preservice mathematics teachers and math majors. Her research focuses are teacher knowledge and mathematical argumentation.

Julie Grasfield Weil is a seventh/eighth-grade English Language Arts teacher who returned to middle school from publishing to help students navigate the turbulent waters of adolescence. She has served on the board of the Sharon Education Foundation for more than a decade. She has been a newspaper columnist for the *Times Advocate* of Sharon and Walpole, MA, for 15 years. She is one of a team of critiquers at *Writer's Infusion*, a Cable TV program that helps aspiring writers hone their craft by constructively critiquing submitted fiction excerpts. She also has written 20 short stories about quintessential adolescent experiences and is currently penning a few YA novels. She received a Bachelor of Arts in Comparative Literature: French and English from Brandeis University and a Master's in Education: Middle School 5-9 from Lesley University. She lives in a suburb of Boston with her husband and two teenage sons.

www.ingramcontent.com/pod-product-compliance
Lightning Source LLC
Chambersburg PA
CBHW022011300426
44117CB00005B/138